Edwin Herbert Hall

Elementary Lessons in Physics; Mechanics, and Light

Edwin Herbert Hall

Elementary Lessons in Physics; Mechanics, and Light

ISBN/EAN: 9783744693370

Printed in Europe, USA, Canada, Australia, Japan

Cover: Foto ©berggeist007 / pixelio.de

More available books at **www.hansebooks.com**

ELEMENTARY

LESSONS IN PHYSICS

*MECHANICS (INCLUDING HYDROSTATICS)
AND LIGHT*

BY

EDWIN H. HALL, Ph.D.

Assistant Professor of Physics in Harvard College

NEW YORK
HENRY HOLT AND COMPANY
1894

INTRODUCTION.

SOME years ago a body of educational leaders declared themselves in favor of teaching physics by means of experiments involving exact measurement and weighing by the pupils in grammar-schools. To the author of this book, well aware of the difficulty of establishing and maintaining a thorough course of quantitative experimental work in academies and high schools, the new proposition did not at first commend itself. Difficulties of various kinds, financial, mechanical, pedagogical, appeared in the way. Indeed, the author at first expressed, somewhat publicly, the opinion that grammar-school physics must be lecture-table physics, an hour or two a week devoted by the teacher to the performance and discussion of simple experiments in the presence of the pupils. He thought, and still thinks, that such a course would be not unprofitable. There are several books describing this kind of work, and their number is rapidly increasing.

But the advocates of the lecture-table method of science-teaching cannot claim for it the disciplinary advantage and the power of bringing the pupil into close quarters with physical facts and laws, that belong to a properly-conducted course of laboratory work by the pupils themselves. If such a laboratory course is readily practicable, grammar-school pupils should have it, for the grammar-school is the popular school, the school in which the great majority of children get the last of their formal education. To what extent, then, is quantitative laboratory work in physics practicable for grammar-schools? The question is here limited to quantitative work, because the author would shrink from the task of laying out a course not mainly quantitative, which would occupy the pupils profitably without making impossible demands upon the time and patience of the teacher.

Quantitative work of a substantial and profitable character is, in the opinion of the author, practicable for grammar-school pupils in the subjects of mechanics (including hydrostatics) and light. Quantitative work in sound, heat, electricity, and magnetism demands apparatus and laboratory facilities that school-boards would at present hesitate to supply to grammar-schools, even if it were certain that pupils of fourteen years could use them to advantage.

This little book has grown out of a course of instruction given by the author, for two years in succession, to teachers conducting, or preparing to conduct, a similar course in the grammar-schools of Cambridge. A large part of the work for pupils described in this book has been actually performed by whole classes in the highest grade of all these schools during the year 1893–94. The success of this "Cambridge experiment" has been, on the whole, gratifying. A brief account of the way in which this new work has been fitted into the school program is given, in an appendix to this book, by Mr. Frederick S. Cutter, who was the first grammar-school master in Cambridge to undertake laboratory teaching in physics.

It is the firm conviction of the author that class laboratory work not accompanied by persistent, energetic, *teaching* is sure to be a failure. We are often told that the favorite method of the elder Agassiz with a new pupil was to set him to gaze in solitude at a single fish for two or three days. Those who would make this the model for science-teaching in general forget that pure observation of numerous, minute, varied details plays a much more important part in natural history than in physics. The teacher of physics who would produce good and lasting results must see to it not merely that the laboratory work shall be carefully done, but that the proper lessons shall be drawn from it and the proper applications made. In fact, the young pupil should give as much time to the study of physics in the lecture- or recitation-room as in the laboratory proper. The *Suggestions for the Lecture-room* given in this book are sometimes very full, but in general they will have to be supplemented by hints from other books or from the teacher's own experience. One or two text-books of the high-school grade, and if possible some book of the college grade, Barker, Deschanel, or Ganot, for instance, should be at the service of the teacher.

The laboratory *Exercises* of this course cover about one third of the ground covered by the laboratory *Exercises* of Hall and Bergen's Text-book of Physics, and most of them are in close correspondence with the work in physics recommended by the Report of the famous "Committee of Ten." It is the hope of the author that the use of this book in the last year of the Grammar School, or the first year of the High School, course will remove much of the difficulty now found by some schools in condensing all of the laboratory work in physics into one year of the High School. The discontinuity thus introduced into the study of physics, a break of two or three years between the study of mechanics and optics, on the one hand, and heat, sound, and electricity and magnetism on the other, is entirely reasonable, in view of the much greater expense and experimental difficulty of laboratory work in these latter subjects.

The course is intended to run through the year and to

occupy the pupil two school-periods, each forty minutes long if possible, per week; one usually in the laboratory, and the other in the lecture- or recitation-room. The number of *Exercises* is only twenty-seven, much less than the number of school-weeks in the year, but some of them may prove to be too long for a single school-period, and teachers will welcome an occasional opportunity for repetition or review. The Cambridge Grammar-schools have given only one school-period per week to the laboratory work, and have, therefore, not been able to do all the *Exercises* in one year.

Although much of whatever is new in this book has originated with the author, many valuable suggestions have come to him from teachers and from makers of apparatus. Perhaps the most striking innovation of the book is a method of measuring the index of refraction of liquids by means of an extremely simple and inexpensive apparatus which yields very satisfactory results.

The book follows, as a rule, the method of leading up to the statement of laws by means of carefully-chosen experiments, rather than the opposite one of giving experiments as illustrations or proofs of laws already stated. It can hardly be said for the former method that it teaches the art of making discoveries,—that art is as difficult to teach as the art of getting rich,—but it has a tendency to keep the pupil in a more active, self-dependent state of mind than the latter method, and in particular it prevents in a large measure that state of bias, or preconception, in the performance of experiments, which is so dangerous not merely to accuracy of observation but to mental rectitude. On the other hand, the teacher using the method of this book must not allow his pupils to think that their experiments, even when most satisfactory, really *demonstrate* the rigid accuracy of any numerical law,—the law of a balanced lever, for instance. He should ask of them, " What law do your experiments *indicate* as true?" and after their answer he should tell them whether their inference is or is not in accordance with the opinion held by those best qualified to judge of the matter in question.

Realizing that this book will naturally be used by teachers little accustomed to physical manipulations or the construction of physical apparatus, the author has taken especial pains in the description of laboratory operations, and has endeavored to give, in Appendix A, complete, detailed, lists of all those articles used in the laboratory exercises and the lecture-room experiments, which are not easily procurable by the teacher. A number of firms, mentioned by name in that Appendix, have undertaken to supply the apparatus described in these lists at reasonable prices. A table like those used in the Cambridge Grammar-schools is described in the same Appendix. Thus the *mechanical* . difficulties of undertaking the course described in this book are reduced to a minimum.

In the Cambridge Grammar-schools the classes, whatever their size, have usually been divided into sections of sixteen or less for laboratory work, and as only one section in a school has worked in the laboratory at one time, only sixteen sets of the pupils' apparatus have been supplied to each school. These sixteen sets can now be furnished for about $80.00 or $90.00. The laboratory for a section of sixteen requires two substantial tables, which may cost $45.00 or $50.00. The teacher's list of apparatus and certain miscellaneous supplies will cost about $30.00. Much, and if necessary all, of the pupils' apparatus can be on shelves under the laboratory tables. Making no allowance for apparatus-cases, and assuming the school-building to have an available room fifteen feet by twenty, or larger, well lighted and supplied with a cold-water tap and a sink, one may estimate the cost of establishing this course fully in any grammar-school at $200.00, a considerable margin of this estimate being intended to cover contingencies not specifically foreseen.

Whatever the merits or demerits of the course laid down in this book, its success in any particular case will depend largely upon local conditions. The author can ask for it no more favorable trial than the good-will of the school authorities and the zeal and ability of the teachers have given it at Cambridge.

TITLES OF THE LABORATORY EXERCISES.

CHAPTER I.

MEASUREMENT OF DISTANCE, AREA, AND VOLUME.

· CHAPTER II.

DENSITY AND SPECIFIC GRAVITY: FLUID PRESSURE.

CHAPTER III.

THE LEVER.

CHAPTER IV.

THREE FORCES WORKING THROUGH ONE POINT.

CHAPTER V.

FRICTION.

CHAPTER VI.

THE PENDULUM.

CHAPTER VII.

LIGHT: REFLECTION.

CHAPTER VIII.

LIGHT: REFRACTION.

PHYSICAL EXPERIMENTS.

CHAPTER I.

MEASUREMENT OF DISTANCE, AREA, AND VOLUME.

Suggestion for the Teacher as to Preparation for Exercise 1.

THE line to be measured may be along the edge of a table (or sheet of paper) from one fine scratch to another, a distance of about 15 inches. It is a great convenience to have all the pupils measure equal distances; accordingly, the teacher is advised to lay off these distances by some method like the following: A carpenter's square is placed along the edge of the table as in Fig. 1, and while it is held firmly in place a fine light scratch is made with the point of a sharp knife-blade at right angles with

FIG. 1.

the edge of the table at the points *a* and *b*. The distance from *a* to *b* is the one to be measured by the pupil. The first-described method of using the measuring-stick in the following Exercise is not a good method, but it is one that many will use if they are not properly instructed. The second method is a good one, and the two are here brought together in order that the pupil may see at once the right way and the wrong way to use such an instrument.

Much of the interest and profit of the Exercise will come from the opportunity given each pupil to compare his own work with that of others.

Any piece of apparatus to be used in the Exercises will usually be referred to by the number it bears in the list of apparatus given at the end of the book.

EXERCISE 1.

MEASUREMENT OF A STRAIGHT LINE.

Apparatus: A short measuring-stick (No. 1) and a meter-rod (No. 2).

To each pupil is given a measuring-stick about one-fourth as long as the distance from *a* to *b*. We will suppose that these sticks are made by sawing a meter-rod, graduated to millimeters, into ten equal parts. The saw-cut will usually leave the divisions at the very ends of the sticks imperfect, and these divisions should not be used in the measurements.

Let each pupil measure his distance at least twice carefully, with his measuring-stick laid flat upon the table, the marks upon the stick being thus *horizontal*, and let him write upon the blackboard the results of his two measurements.

Then let each pupil measure his distance twice again, this time placing his measuring-stick upon its edge, so that the marks upon it will be vertical, making a light, fine mark upon the table with a sharp pencil to *set* the stick by, whenever it is moved forward a length. These new measurements are also to be placed upon the blackboard under the first ones.

Finally let each pupil measure his whole distance at once with his meter-rod and write this last measurement with the others.

To judge of the accuracy of a set of measurements it is not enough to know how much these differ among themselves, for the importance of the difference usually depends upon the ratio which the difference bears to the whole quantity measured. A thousandth part of an inch might be a very serious difference to a watchmaker in the measurement of some small cylinder, while a difference of several inches in the measurement from one mile-post to another would be of but little consequence. The pupil should therefore form the habit of comparing his errors, or the differences of his measurements, with the whole quantity that he had to measure.

Let us suppose, for instance, that in this Exercise the measurements made by one pupil are 37.30 cm., 37.00 cm., and 37.10 cm. The greatest difference is found between the first and second. It is 0.3 cm., and its ratio to 37.15 cm., which is midway between 37.30 cm. and 37.00 cm., is 0.0081. We see, then, that the difference between the two measurements of the line is about *eight one-thousandths,* not quite one per cent, of the length of the line.

Each pupil should make a similar calculation from his own measurements in Exercise 1.

Suggestions for the Lecture-room.

The importance of having definite units of length, of weight, etc., so that any man in dealing with his neighbor may know just how much is meant by the words *foot, pound,* and the like, is so great that in all civilized countries the exact meaning of such words is fixed by law, and very great care is taken to make and preserve government *standards,* as they are called, standard yard-sticks, standard pound-weights, for instance, with which as patterns the measuring instruments used in business are compared and tested.

Interesting accounts of the foot, the yard, the meter, etc., can be found in almost any encyclopedia.

Meter-rods for school use are in many cases marked off in inches on one side. With the information given by such a rod the class can find how many centimeters are equal to one inch. This number carried to two places of decimals is accurate enough for most purposes.

EXERCISE 2.

THE LINES OF THE RIGHT TRIANGLE AND THE CIRCLE.

Apparatus: A 30-cm. measuring-stick (No. 3). A sheet of paper upon which is drawn carefully a right triangle no side of which is less than 10 cm. long. (No two pupils should use exactly similar triangles.) A cylinder of wood 4 or 5 cm. in diameter (No. 4). A narrow straight-edged strip of thin paper.

PART 1.—MEASUREMENT OF THE SIDES OF A RIGHT TRIANGLE. —Let each pupil measure very carefully all the sides of his triangle, not being content to read to the nearest 0.1 cm., but striving to note and measure 0.05 cm. distances, if he can do so without hurting his eyes.

After the measurements are made square the length of each side and compare the greatest square with the sum of the other two squares. The conclusion drawn from this comparison must not be extended to triangles which are not right-angled.

PART 2.—MEASUREMENT OF THE CIRCUMFERENCE AND DIAMETER OF A CIRCLE.—Measure carefully the diameter of one end of the cylinder. Then wrap the strip of paper around the curved surface of the cylinder at the same end, and mark upon the edge of the strip the point where the second winding of the paper begins to overlap the first. Then unfold the paper and measure upon it that distance which extended once around the cylinder. Then divide this distance, which of course is equal to the circumference of the circle, by the length of the diameter. The ratio thus obtained is one which it is important to know, although we shall not have much occasion to use it in this book. Mathematicians, physicists, and engineers use it so much that they have a particular sign, π, to denote it.

This sign is a Greek letter and is called *pē* by students of Greek, but when used as just described it is often called *pie* to distinguish it from *p*.

Suggestions for the Lecture-room.

The measurements of Exercise 2 may be discussed somewhat as follows: The square of the longest side of the triangle is found by one pupil to be 404.01 and the sum of the squares of the other two sides 406.05. If the two short sides were measured correctly, how large an error in the measurement of the longest side would cause the disagreement here found? The long side was measured as 20.10 cm. If it had been called 20.20 cm., its square would have been 408.04, which is about as much too large as the square actually found is too small. If the distance had been measured as 20.15 cm., the square would have been 406.02, a quantity very close indeed to the sum of the other two squares. If, therefore, the original error lay entirely in the measurement of the longest side, this error must have been very nearly 0.05 cm. Of course the error may have been made in measuring the other sides, or in drawing the triangle, or in all parts of the work. An error which mistakes 20.15 for 20.10, or 201.5 for 201.0, or 2015 for 2010, is called in each case an error of 5 parts in 2015, or 1 part in 403, or an error of about $\frac{1}{4}$ per cent (see remarks following Exercise 1).

Question for the Class.

In the case of the circle, which would make the greater difference in the result (circumference ÷ diameter), an error of 0.05 cm. in the measurement of the diameter or an error of 0.10 cm. in the measurement of the circumference?

SURFACE.

Thus far we have been measuring the length of lines. To measure a line, as we see, is merely to find out by trial that it is so many centimeters or inches long. A line 10.6 cm. long is one that could be divided into ten full centimeters and six tenths of another centimeter. We here call the centimeter our *unit* of length.

If we have to measure a *surface*, the whole table-top, for instance, our task is to find the number of square centimeters, or square inches, or square feet, that would be required to cover it, or that it would make if it were cut up without waste into squares. In this case the square centimeter, or square inch, or whatever square we choose to take, is the *unit* of surface. We might set about to measure surfaces by actually placing a little square, a square centimeter, for instance, on the given surface, marking a line close around it, then moving it to a new place, marking around it, and so on till we had marked off the whole surface into little squares, with perhaps some fractions of squares. But this is not the common or the best way of measuring surfaces. The common way is to measure the length of certain lines on the surface and from the lengths of these lines *calculate* the extent of the surface. If the surface is in the form of a rectangle, like Fig. 2, it is plain that we have merely to multiply the number of units, centimeters let us say, in the length by the number of centimeters in the width, and the result,

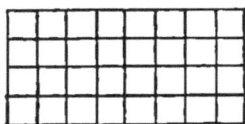

Fig. 2.

$8 \times 4 = 32$ in this figure, is the number of square centimeters into which the surface can be divided. This is called the *extent* or *area* of the surface.

In the next two Exercises we shall undertake to find rules for the measurement of surfaces not quite so simple in shape as the rectangle shown in Fig. 2. These will be parallelograms and triangles.

A *parallelogram* is a flat figure bounded by four straight

Fig. 3.

lines, each line being parallel to the line opposite. Thus

A and *B* in Fig. 3 are parallelograms. *A* is what we have just called a rectangle, and we have seen how to find the area of any rectangle, but *B* is not quite so simple at first sight. A parallelogram like *B*, which contains no right angle, is called an *oblique* parallelogram.

EXERCISE 3.

AREA OF AN OBLIQUE PARALLELOGRAM.

Apparatus: The 30-cm. measuring-stick (No. 3). An oblique parallelogram of paper about 20 cm. long and 10 cm. wide. (One of the straight-edged rulers (No. 23) may prove useful in this Exercise.)

Draw upon the paper figure a line like *c* in Fig. 4, taking care to make a right angle with the top line and the bottom line, and then cut or tear the paper along the line *c*. Take the small piece thus removed and join it to the larger piece, in such a way as to make a figure that you know how to measure. Measure the length and width of the figure thus formed and calculate the extent of its surface.

FIG. 4.

Then put the two pieces together as they were at first and ask yourself whether you could not, if another parallelogram were given you, find the extent of its surface without cutting it.

Suggestions for the Lecture-room.

Have the pupils *estimate* without measurement the length and width of some visible and convenient rectangles, a book-cover, a table-top, a window, etc., and calculate the areas from these estimated dimensions. Then give them the true dimensions and let them calculate the true areas.

EXERCISE 4.

AREA OF PLANE TRIANGLES.

Apparatus: Measuring-stick (No. 3). A right triangle and an oblique triangle of thin paper.

Take the right triangle, represented by Fig. 5, and draw upon it a line *d* parallel to the base line *cb*, beginning at a point midway between *a* and *b*. Fold the paper along the line *d* and then tear it along the same line. See whether you can put the two pieces together in the form of a rectangle. If so, measure the length and width of this rectangle, then put the pieces together again in their original position.

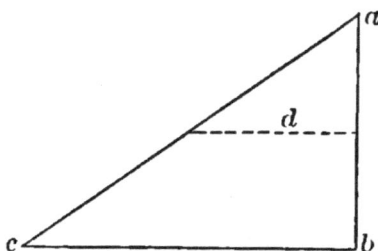

We shall call the side *cb* the *base* of the triangle, and the side *ab* the *height*, or *altitude*. Can you now frame a rule by means of which you can find the area of a right triangle without cutting it?

FIG. 5.

Now take the other triangle *abc*, Fig. 6, and draw a line *d*

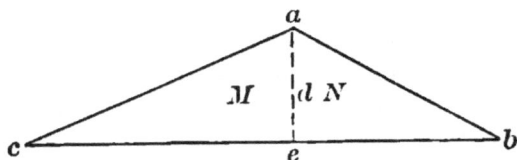

FIG. 6.

from the angle *a* in such a direction that it will make right angles with *cb* at *e*. We will call *cb* the base of the whole triangle and *d* its height. This line *d* has now divided the original triangle into two, *M* and *N*, each of which is a right triangle. You have already learned how to find the area of a right triangle. You can say at once the

area of *M* = length of . . . × . . . length of . . .
"   " *N* =   "   "   . . . × . . .   "   "   . . .

area of whole triangle = . . . × . . . length of . . .

You can now frame a rule for finding the area of any triangle.

Suggestions for the Lecture-room.

Any plane figure bounded by straight lines, Fig. 7, for instance, can evidently be divided into triangles and each of these triangles can be measured. The sum of the areas of the triangles will be the area of the whole figure. We see, then, how to measure the area of any plane figure bounded by any number of straight lines.

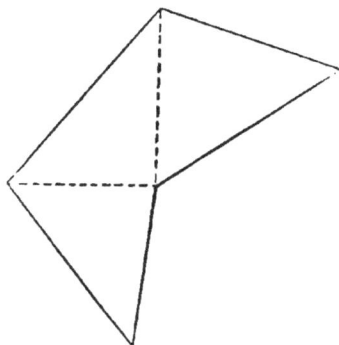

Fig. 7.

VOLUME.

We have now to speak of the measurement of *volume*. The *unit* of volume may be the cubic centimeter, or the cubic inch, or the cubic foot, etc. We shall generally use the cubic centimeter as our unit.

We mean, then, by the *volume of a body* the number of cubic centimeters that could be made of that body if it were cut up without waste, as one might cut up a large piece of clay or putty. In the case of a body whose surface is made up of rectangles, a brick, for instance, it is easy to see how the volume may be calculated, if we know the length and the width and the thickness. We have *volume = length × width × thickness.* If the body is of less regular shape, like an ordinary stone or a lump of coal, it is not so easy to calculate its volume from measurements of length, width, and thickness. There is, however, a very easy way of finding the volume of such a body by the use of water, as will presently be seen. It is easy to find the volume of a quantity of water in several ways. One way is to pour the water into a rectangular box. Then we can measure its length and width and depth and calculate its volume. Another way is to pour it into a glass measuring-dish having marks upon it to tell the number of cubic centimeters required to fill it to certain depths. Another method is to weigh the water, for it is known that one cubic centimeter of water weighs one *gram*. Indeed this is the *definition* of one gram, the weight of a cubic centimeter of water.* If the balance which we use for weighing reads in ounces instead of grams, we shall have to remember that *1 oz. = about 28.3 gm.*, so that 1 oz. of water will be 28.3 cubic centimeters. We shall commonly find the volume of a body of water by *weighing*.

We will now try the water method of finding the volume of a body, a rectangular solid. We will find its volume by the water method and also by direct measurement and calculation, and then see how well the two results agree. This will test the water method, and if we find it to work well, we can use it with irregular solids which we cannot measure directly.

EXERCISE 5.

VOLUME OF A RECTANGULAR BODY BY DISPLACEMENT OF WATER.

Apparatus: A brass can (No. 5) called *C* in Fig. 8. A small catch-bucket (No. 6) called *p* in Fig. 8. A spring-balance (No. 7). A rectangular block of wood (No. 8) so loaded as to sink in water.

Closing the overflow tube *t* of the can *C*, pour water into *C* until it is filled nearly to the brim. Then open the tube and let all the water flow out that will do so, catching it in the small can *p*. The large can should rest steadily upon the table, but the small one is better held in the hand when the flow begins, otherwise some water may be spilled. About one minute should be allowed for the flow and the dripping which follows.

Throw away all the water thus caught in *p*, and then weigh *p* on the spring-balance to the nearest gram or the nearest twen-

---

* To be exact one must add *at 4° of the centigrade scale of tempera-ture*. For the purpose of this book such exactness is unnecessary.

tieth of an ounce, according to the graduation of the balance.*
Then, closing the tube *t* as before, lower into the can *C* the
wooden block until it rests upon the bottom. Then, or sooner if

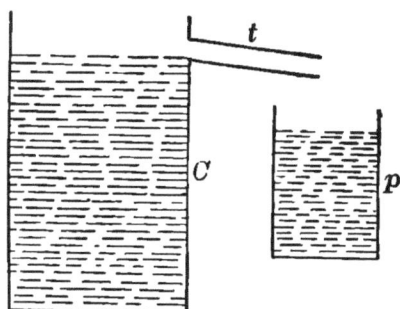

FIG. 8.

the can *C* seems likely to be overflowed, open the tube *t*, and as
before catch the water that runs out in the small can *p*. The
water, Fig. 9, now stands just as high in *C* as it did just before the

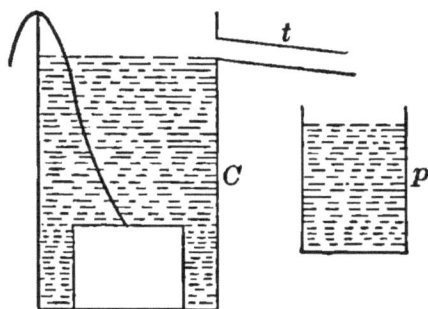

FIG. 9.

block was put into it. The block has crowded out into the can *p*
just its own bulk of water. If, then, we can find the volume of
the water that the block drove over into *p*, we have the volume
of the block itself.

---

* Ordinary small spring-balances now in the market are marked
off in ½-ounce divisions, which are about ¼ inch long. The pupil
will learn to *estimate* the position of the pointer when it falls between
two lines, so as to read to about $\frac{1}{20}$ of an ounce.

Weigh p and the water it contains.

Weight of small can and water	$=$	....
" " " " **empty**	$=$	....
" " **water alone**	$=$	....

If the weight as thus found is in grams, it is equal to the number of cubic centimeters in the block. If the weight as thus found is in ounces, we must multiply the number of ounces by 28.3 in order to find the number of cubic centimeters in the block.

Now measure carefully the length, width, and thickness of the block and calculate the number of cubic centimeters it contains from these measurements.

(Experiments for finding the volumes of irregular bodies by the water method may well be postponed till the next Exercise, which would otherwise be a very brief one. Potatoes, stones, lumps of coal, etc., of suitable size may be used for these further experiments.)

Suggestions for the Lecture-room.

Give simple questions and problems, to be answered at once, upon the preceding Exercises.

Draw a line 10 inches long on the blackboard, divide it by a cross-line at any point, and ask the class to estimate the distance from either end to this cross-line.

Show how to test the correctness of the spring-balances by hanging known weights upon them, 2 ounces, 4 ounces, 8 ounces, for instance.

CHAPTER II.

DENSITY AND SPECIFIC GRAVITY: FLUID PRESSURE.

EXERCISE 6.

WEIGHT OF UNIT VOLUME OF A SUBSTANCE.

Apparatus: A block of wood (No. 9). A spring-balance (No. 7). A measuring-stick (No. 3). Thread for suspending the block.

Find the weight of the block in grams and also in ounces.

Measure the length of each of the four edges which are parallel to the grain of the wood, take the average of these measurements and call it the *length* of the block.

Measure the length of each of the four long edges which are crosswise to the grain of the wood, and call the average of these four measurements the *width* of the block.

Measure the length of each of the four short edges and call the average of these four measurements the *thickness* of the block.

The weight in ounces is to be turned into pounds.

From the length, width, and thickness in centimeters the length, width, and thickness in *feet* may be found by the rule that 1 ft. = 30.5 cm., but it is shorter to find the volume in feet from the volume in cubic centimeters by the rule that 1 cu. ft. = 28300 cu. cm.

Calculate, 1st, how many grams, or what part of a gram, 1 cu. cm. of the block weighs; 2d, how many pounds, or what part of a pound, 1 cu. ft. of such wood weighs.

Suggestions for the Lecture-room.

The weight of unit volume of a substance is called the *Density* of the substance.

We have found that the density of a body in gram and centimeter units is not the same as the density of the same body in pound and foot units.

The weight of 1 cu. cm. of water is 1 gram; but the

weight of 1 cu. ft. of water is about 62.4 pounds. This is a useful fact to remember.

What ratio has the class found between the density of wood in pounds and feet and its density in grams and centimeters ?

If we know the density of a substance we can calculate the weight of any volume of that substance. Engineers and other scientific men often have to find by this method the weight of objects which it would be inconvenient to weigh. The weights of buildings and bridges, for instance, are found in this way. Books used by scientific men contain tables giving the densities of many different substances.

Often we find it useful to know the ratio between the *weight of a body* and the *weight of an equal bulk of water*, and we shall have soon a number of Exercises showing how this ratio may be found. Before we come to these we shall need to go through one or two preliminary Exercises to make us better acquainted with the force exerted by water upon bodies floating or immersed in it.

Before going farther we need to think carefully about the meaning of the word *weight*, which we have already used a number of times and shall have to use very often. The word has two meanings.

Sometimes when we speak of the weight of a body we mean the *amount* of the body, as when we speak of 10 lbs. of butter or 100 lbs. of iron.

At other times we mean by the weight of a body the amount of the earth's downward pull upon that body, as shown by the spring-balance, for instance.

It is somewhat hard to remember this distinction, because the *units* in which we tell the amount of a *body* have the same name as the units in which we tell the pull which the earth exerts upon the body. For instance, we say that the earth exerts a pull, or *force*, of 5 lbs. upon

5 lbs. of wood, or 5 lbs. of coal, or anything which consists of, or *is*, 5 lbs. of substance.

Often when we use the word *weight* it makes no difference which of its two meanings we have in mind, but sometimes it does make a difference. Thus, when we put a body under water, as we shall do in the next Exercise, and say that it loses in *apparent weight* in going from air to water, we do not mean that there appears to be any less of the *body* in water than there was in air. We mean that it requires a smaller pull of the spring-balance to keep the body from sinking in water than it does to keep it from sinking in air.

EXERCISE 7.

LIFTING EFFECT OF WATER UPON A BODY ENTIRELY IMMERSED IN IT.

Apparatus: Overflow-can (No. 5). Catch-bucket (No. 6). Spring-balance (No. 7). Loaded block (No. 8). Thread.

Fill the can and let it overflow and drip for one minute as in Exercise 5. Catch this overflow in the small bucket and throw it away. Then weigh the empty bucket in grams.

Weigh the block in grams before immersing it in the water.

Lower the block, still suspended from the balance, into the overflow-can till it is entirely covered, catching the overflow and saving it.

Weigh the block in the water, the balance being entirely above the water.

Weigh the bucket with the overflowed water.

Subtract the (apparent) weight of the block in water from its weight in air, and call the difference the *loss of weight of the block in water*, or the *buoyant force exerted upon the block by the water*.

Find weight of the water in the small bucket, and compare this with the loss of weight of the block in water.

If there is time, make a similar experiment with other bodies.

The law illustrated in this Exercise is called from its discoverer the *law*, or *principle*, of *Archimedes*. (See any encyclopedia for an account of Archimedes.)

Suggestions for the Lecture-room.

Fill the gallon glass jar (No. 10) with water to a level about one inch from the top. Close the smaller end of a student lamp-chimney tight with a good cork stopper. Make the pressure-gauge (No. I.) ready for use by the following operation, having first put on a fresh rubber diaphragm if necessary: Release the glass tube from the rubber tube and wet the whole length of the glass tube inside with water, leaving within it, about one inch from one end, a column of water about one-half inch long to serve as an index.* Hold the gauge itself under water for a little time before reconnecting the glass tube with the rubber tube, in order to allow the air within the gauge to come to the temperature of the water. Now push the gauge down into the jar and raise and lower it repeatedly in the water, keeping the glass tube with the water-index horizontal, and let the class determine from the movements of this index whether the pressure of the water against the rubber diaphragm increases or decreases when the gauge is pushed deeper in the water.

Rest the bottom of the wooden pillar of the gauge upon the bottom of the jar, and, still keeping the glass tube horizontal, turn the upper pulley so that by means of the rubber band the lower pulley will be turned and the rubber diaphragm will face downward sidewise and upward in succession, its centre remaining practically unchanged in position. Let the class determine by watching the water-index whether the pressure upon the rubber diaphragm is any greater when it faces upward than when it faces downward or sidewise.

Push the closed end of the lamp-chimney down into the

---

* It may be necessary to use water colored by some aniline dye before a large class.

water till it is near the bottom of the jar. Move the gauge-face about, without changing its level, so as to bring it under this closed end. Move it now out of and now into this position, thus changing the depth of water immediately above it from one-half inch or less to several inches. Let the class determine by watching the index whether such changes of position, without change of *level*, make any difference in the pressure against the gauge-face.

We shall make considerable use farther on of the facts brought out by these experiments. Just here we can see that they explain, at least in a general way, why a body immersed in water weighs, or appears to weigh, less than when in the air. For we see that there is an upward pressure of the water against the under side of the body, and that this upward pressure is greater than the downward pressure against the upper side of the body.

Having seen that there is greater pressure on low levels than on high levels in water, we may well ask whether this greater pressure crowds the particles of water closer together on the low levels, thus making the water *denser* than on high levels. In fact there is an effect of this kind, but it is so slight that we need take no account of it in any ordinary case. It is very difficult to compress water much.

EXPERIMENT.

Fill a bottle with water and close it with a rubber stopper, leaving one hole through it. Then, holding the stopper firmly in place, try to push into the hole a solid brass rod of a size to fit rather closely. (App. No. II.)

We have not made, and cannot well make with the *gauge* just used, any accurate measurement of the rate at which pressure changes with change of level in water. The fact is, however, that if we place a surface of 1 sq. cm. horizontal at any depth in water the column of water just above it is resting upon the given surface.* If we carry the given surface down 1 cm. farther, we now have resting upon it a load somewhat greater than before, greater by the weight of the additional 1 cu. cm. of water which is now above it. As 1 cu. cm. of water weighs 1 gm., the pressure upon a surface of 1 sq. cm. changes by 1 gm. for each 1 cm. change of level in the water.

EXERCISE 8.

WEIGHT OF WATER DISPLACED BY A FLOATING BODY.

Apparatus: The same as in the preceding Exercise, with the exception of the sinking body, which is here replaced by one that floats (No. 4).

Weigh the cylinder, in grams, in air. Find, in grams, the weight of water which it displaces from the overflow-can. . Compare these two weights.

Suggestions for the Lecture-room.

We have made some experiments with liquid-pressure. We must now begin to learn something about air-pressure, which in many practical matters of every-day life has a very important connection with water-pressure. We will at the start repeat in a slightly varied form a famous experiment first made by Torricelli, an Italian, about 250 years ago.

EXPERIMENTS.

Take two pieces of strong glass tubing about 0.7 cm. in inside diameter, one of them, about 1 m. long, closed at one end, the other, about 20 cm. long, open at both ends, and connect them by means of a thick-walled piece of rubber

---

* The pressure upon the given surface may be greater than the weight of the column of water resting upon it, for there may be, and usually is, a downward pressure of air or something else upon the top of the water-column.

tubing about 25 cm. long. The rubber tube should fit tight upon the glass tubes, and for greater security should be fastened on by means of wire or string.

Holding the tubes thus connected (App. No. III.) by the free end of the short glass tube, the closed end of the long glass tube hanging down, pour mercury by means of a small funnel of glass or paper into the tubes, tapping or shaking them occasionally to dislodge air-bubbles, until the top of the mercury-column reaches the rubber tube. Then gently raise the closed end of the long glass tube until this tube points straight upward (Fig. 10), meanwhile holding the other glass tube upright and taking care that no mercury is spilled.

During the latter part of this operation it will be noticed that the mercury begins to fall away from the closed end of the long glass tube, and finally several inches of this tube will be apparently empty. (Really this space contains a very little air, from the bubbles that were in the mercury-column before it was inverted, but so little

FIG. 10.

that we may at present disregard it and consider the space above the mercury as empty. Such a space is called a *vacuum*, from a Latin word meaning *empty*.) But the mercury continues to stand very much higher in the long glass tube than in the short one. It was known before the time of Torricelli that if air was drawn out from a tube the lower end of which rested in water, the water would rise in the tube, but the true reason for this was not known. Torricelli maintained, and Pascal, a Frenchman, showed by experimenting at different heights in the air, that the pressure of the atmosphere, due to its weight, accounted for the rise of liquids in a vacuum. We have only to think of the fact that the air, although its density is very small compared with that of water, has, because of its great

quantity, a great weight, and we see that the air, pressing upon the mercury surface in the shorter tube, balances the column of mercury in the long tube.

By measuring the difference in height of the two mercury surfaces we can get a measure of the atmospheric pressure. We find that the atmospheric pressure is about as great upon the surface of the earth as would be the pressure of a layer of mercury 76 cm. deep, or a layer of water about 10.3 m. deep, over the whole earth. The pressure per square centimeter at any given part of the earth's surface varies somewhat from day to day and even from hour to hour.

If we fasten the apparatus that has just been used to a suitable support it will serve as a fairly good barometer.

Air-pressure, like liquid-pressure, is at any given point equal in all directions, if the air is at rest.

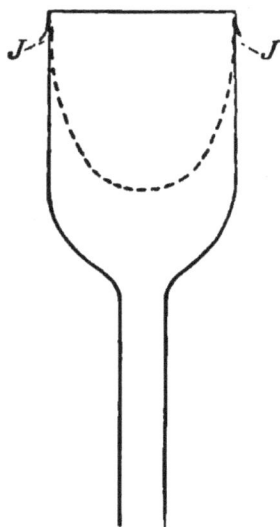

Fig. 11.

Take a strong thistle-tube (No. IV.) of the shape shown in Fig. 11 and tie a piece of thick sheet rubber across the mouth, which may be about 1 inch in diameter. Make the covering air-tight by means of some cement, melted beeswax and rosin, for instance, poured in at the joint *J.* Connect this thistle-tube by means of a thick-walled rubber tube to an air-pump (No. V.), and exhaust the air. The rubber cap, not being supported by air-pressure beneath, will now be pushed down by the atmospheric pressure into a deep cup-shape. Pinch the rubber tube so that no air shall leak back into the thistle-tube, and then turn the mouth of the latter in all directions, sidewise, downward, and oblique. Observe whether the form of the rubber cup changes during this operation, as it would do if the pressure upon it changed.

We should find by proper experiments that in air at rest, as in water at rest, pressure is equally great at all points on the same level. We should find, also, that the air-pressure diminishes with increase of height from the earth's surface, but, as the density of air is very little compared with that of water, it requires a considerable change of level to make much difference in the air-pressure.

Substances which, like water and air, press equally in all directions at a given point and are easily changed in shape are called *fluids*, that is, substances that can *flow*. Other substances, like wood, iron, stone, etc., which do not *flow* from one shape to another are called *solids*.

Fluids are divided into two classes: *liquids* and *gases*. Water, oil, milk, kerosene, etc., are liquids. Air is a mixture of several gases. Liquids are much heavier than gases, in most cases. Most liquids are easily *seen*. Most gases are practically invisible. But perhaps the most striking difference between liquids and gases is a difference in compressibility. We have seen that it is difficult to compress water much, but it is very easy to compress air.

Take the bent glass tube (No. VI.), closed at one end, and pour into it a little mercury, enough to fill the bend. At first the mercury will stand a little higher in the long arm, but by tipping the tube and letting out a little of the air imprisoned in the short arm the level can be made nearly the same in both arms, as in Fig. 12. Now measure the length of the imprisoned air-column, and write it under the letter $V*$ on the blackboard.

V.	P.	$V \times P$.
....	....	....
....	....	....
....	....	....
....	....	......

The pressure upon this air is now, if the mercury level is the same in both arms, equal to that upon the unimprisoned air. It is as great a pressure as would be exerted by the weight of a column of mercury as tall as that in the barometer (Fig. 10). Take, then, a reading of this barometer and record this reading under the letter P.

FIG. 12.   Pour in more mercury till the difference of level in the two arms is about 20 cm., then measure again the length of the inclosed air-column. Record this length under V, and record under P the present difference of mercury level *plus* the height of the barometer column.

Proceed by stages in this way till the volume of the inclosed air-column is about one half what it was at first. Multiply each number under V by the corresponding number under P, and write the products in the column headed $V \times P$. An examination of this last column will probably indicate a very simple law connecting pressure and volume in the case of any given body of air. This law is important, and should be remembered by the pupil. It is sometimes called Boyle's law and sometimes Mariotte's law. We shall call it by the shorter name, *Boyle's law.*

---

* The *length* of the air-column is the same as its *volume*, if we take for our unit of volume the space contained in unit length of the tube.

EXERCISE 9.

RATIO BETWEEN THE WEIGHT OF A SOLID BODY AND THE WEIGHT OF AN EQUAL BULK OF WATER.

Apparatus : The spring-balance (No. 7). The gallon jar (No. 10) nearly filled with water. A lump of sulphur (No. 11). Thread.

Weigh the sulphur in air ; then in water.

We know from Exercise 7 that a body immersed in water loses in apparent weight an amount equal to the weight of the water whose place it has taken. It is easy, therefore, to get from the two weighings just made the ratio which we have undertaken to find in this exercise.

This ratio is called the *specific gravity* of sulphur as compared with water. We shall have a number of exercises for finding specific gravities by other methods. (*Gravity* comes from a Latin word *gravis*, meaning *heavy*. *Specific* means *distinct*, or *particular*. The *specific gravity* of a body is its *particular heaviness*—the heaviness which distinguishes this body from other bodies of equal size.)

If time permits, find in this Exercise, by the same method that was used for the sulphur, the specific gravity of other solids that will sink in water—glass, coal, etc.

Suggestions for the Lecture-room.

QUESTIONS.

A cubical box, 10 cm. along each edge, has extending from its top, as in Fig. 13, a tube 15 cm. tall and 1 sq. cm. in cross-section (inside). If the box, but not the tube, is full of water, how great is the water-pressure on the whole of the bottom ?

If the tube as well as the box is full of water, how great is the pressure upon that one sq. cm. of the bottom which lies just beneath the tube ?

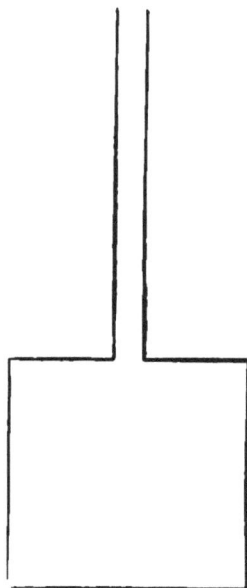

FIG. 13.

Is the pressure equally great, per sq. cm., at other parts of the bottom?

How much is the total pressure now on the bottom of the box?

How great is the pressure per sq. cm. at the top of the box just at the bottom of the tube?

How great is the total *upward* pressure of the water against the top of the box?

(Disregard the atmospheric pressure upon the top of the water-column in all these questions at first. Afterward call this atmospheric pressure 1000 gm. per sq. cm., and ask the same questions as before.)

EXPERIMENTS.

The preceding questions bring out the fact that by exerting pressure in a small tube connected with a large vessel, both being filled with water, one can increase correspondingly the pressure throughout the vessel. The following experiment will show that similar effects can be produced with air-pressure : Take a common rubber football and blow air into it till it is about half filled, connecting a rubber tube with the key for greater convenience in blowing (App. No. VII.). Then rest one end of a board,

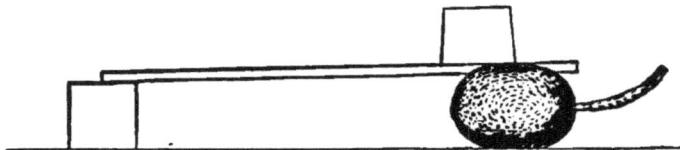

Fig. 14.

as in Fig. 14, on the football and the other end upon a box or block of about the same height. Then place a weight of 25 lbs. or more on the board nearly over the ball, holding the rubber tube attached to the key in such a way that the air cannot escape from the ball. Then blow through the tube into the ball and observe that you can in this way lift the weight.

The experiment illustrates the operation of the *hydrostatic press*, a machine in which a very great force is obtained, for lifting or compressing bodies, by pumping water through a small tube into a large cylinder, one end of which is closed by a movable stopper called a *piston*.

Take again the pressure-gauge and the accompanying apparatus used in the experiments following Exercise 7. Fill the lamp-chimney with water, and then, holding a card across the open end, invert the chimney, lower the end covered by the card into the water, and then remove the card. Most of the water will now remain in the chimney, although its upper end is nine or ten inches above the surface of the water in the jar.

How does the pressure per sq. cm. inside the chimney on a level with the outside water-surface compare with the pressure per sq. cm. at this outer surface, that is, the atmospheric pressure? How, then, will the pressure per sq. cm. at points higher in the chimney compare with the atmospheric pressure? After answering these questions by the aid of what the class already knows about liquid pressure, test the correctness of the answer by means of the gauge. (The gauge as mounted for previous experiment is rather inconvenient for this experiment with the lamp-chimney, and it would be well to detach it from the wooden support if this can be done readily.)

Take a long narrow glass tube open at both ends, and dip one end into a vessel of water. Apply the lips to the other end and draw the water up till the tube is filled. Ask the class to explain the operation. In what sense is the water *drawn* up? (The operation begins with an expansion of the lungs which lessens the air-pressure within them. Then air runs from the place of high pressure, the tube, to the place of low pressure, the lungs. So the air-pressure within the tube is lessened.)

After filling the tube in this way quickly close the top with a finger and then lift the lower end from the water. Take off the finger for an instant, then replace it.

Fill or nearly fill a tumbler or broad-mouthed bottle with water and then cover it with a sheet of thick paper. Hold the paper firmly in place with the hand and invert the tumbler; then take away the hand that holds the paper. As accidents may happen, the tumbler should be held over some large dish.

EXERCISE 10.

SPECIFIC GRAVITY OF A BLOCK OF WOOD BY USE OF A SINKER.

Apparatus: A rectangular block of wood (No. 9). The spring-balance (No. 7). The gallon jar (No. 10) nearly filled with water. A lead sinker (No. 12). Thread.

By definition (Exercise 9) we have

$$\text{Sp. grav. of a body} = \frac{\text{Wt. of the body}}{\text{Wt. of an equal volume of water}}.$$

Exercises 5 and 7 show us that the quantity written below the line in this definition may be expressed in other ways. We may write

Sp. grav. of a body

$$= \frac{\text{Wt. of the body}}{\text{Wt. of water displaced by the body when immersed}}$$

or

$$\text{Sp. grav.} = \frac{\text{Wt. of the body}}{\text{Loss of weight of the body when immersed}}$$

or

$$\text{Sp. grav.} = \frac{\text{Wt. of the body}}{\text{Lifting effect of water upon the body when immersed}}$$

These expressions all mean the same thing, but sometimes one of them is more convenient than the others. In the exercise now before us we shall use the last form.

We have to find two quantities, by experiment: 1st, the *weight of the body ;* 2d, the *lifting effect of water upon it when immersed.*

Weigh the wood in air and record its weight.

Now put the block into water. You see that it floats. To make it stay under water you must *hold it down.* Try this, putting your fingers on the block. In this case, you see, the lifting effect of the water, when the block is wholly beneath its surface, is greater than the weight of the block. We will try to find out how much it is.

We shall use the lead sinker to hold the block under water, and we need to know the weight of the sinker alone under water. Weigh it in this position and record the weight.

Now suspend the block from the balance * and the lead sinker from the thread under the block, and consider how much the two, block and sinker, would weigh in the position shown by Fig. 15, the block out of water and the sinker in water. You can tell this from the weighings already made. Write it down:

Wt. of block in air + *Wt. of sinker in water* = ... + ...

Now lower the block and sinker till both are covered

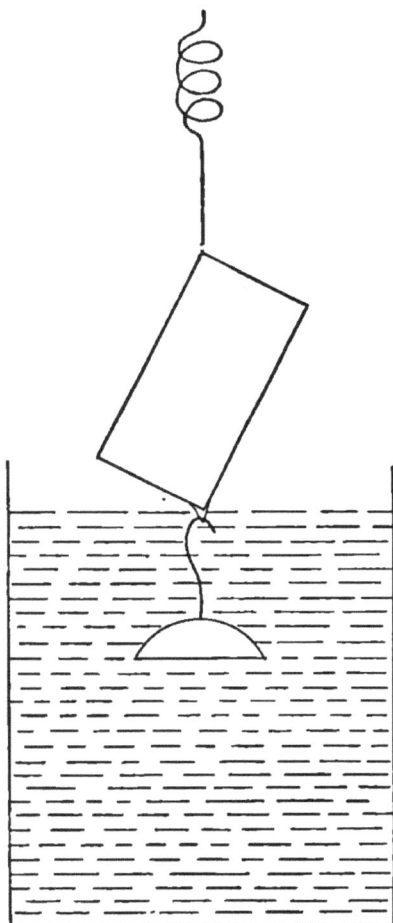

FIG. 15.

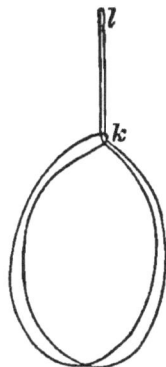

FIG. 16.

* The success of a difficult experiment like this depends greatly upon the care with which the details of the work are thought out by the teacher. The following method of attaching the block to the balance is recommended: Take a thread two feet long and tie the ends together. Then make of it a slip-noose by passing one end, *l* (Fig. 16), through the other end, *k*. The block may then be placed in the noose and the loop *l* slipped upon the hook of the balance, but to prevent slipping when the lead weight is to be suspended from the loop below the block it is well to pass the loop *l twice* through at *k*.

by the water, and weigh the two together in this position and record :

Wt. of block and sinker together in water = ...

Just before the *block* entered the water, the sinker being already in, the weight was . . . . Just as soon as the block also was covered the weight was only . . . . The difference is the *lifting effect of the water upon the block.* We have now all that we need for calculating the specific gravity of the block by means of the formula already given,

$$\text{Sp. grav.} = \frac{\text{Wt. of block}}{\text{Lifting effect of water upon block immersed}}.$$

Suggestions for the Lecture-room.

Review questions on liquid pressure and specific gravity.

PROBLEMS.

(1) A brick-shaped body 20 cm. long, 10 cm. wide, and 5 cm. thick weighs 1500 grams. What is its *density* in gram and centimeter units ?

What would be the weight of an equal bulk of water ?

What, then, is the *specific gravity* of this body ?

(2) A body whose volume is 700 cu. cm. has the density 8 in gram and centimeter units. How much does it weigh ? What is its specific gravity ?

(3) A body 20 ft. long, 10 ft. wide, and 5 ft. thick weighs 93,600 lbs.

What is its density in pound and foot units ?

What would be weight of an equal bulk of water, one cu. ft. of water weighing 62.4 lbs. ? What, then, is the specific gravity of the body ?

(4) A body whose volume is 700 cu. ft. has the density 499.2 in pound and foot units. How much does it weigh ? What is its specific gravity ?

What numerical relation do we find in the preceding

problems between density in gram and centimeter units and specific gravity?

What relation between density in pound and foot units and specific gravity?

EXPERIMENT.

Fig. 17 (App. No. VIII.) shows a bottle closed with a rubber stopper through which two glass tubes, *a* and *b*, open at both ends, extend. To one of the tubes, *a*, is attached a rubber tube, *r*. The bottle and the two glass tubes are full of water.

By applying the lips to the outer end of the tube *r* water can be "drawn" into the mouth. Can this be done when the tube *b* is closed by a finger at the top?

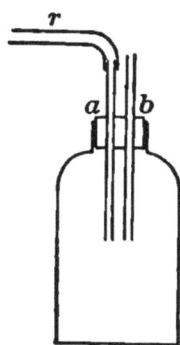

FIG. 17.

EXERCISE 11.

SPECIFIC GRAVITY BY FLOTATION METHOD.

Apparatus: The gallon jar (No. 10) nearly filled with water. A slender wooden cylinder (No. 13). A support for holding this cylinder upright in water (No. 14). A measuring-stick (No. 3).

If a cylinder floated upright with its top just level with the top of the water, we should at once know its specific gravity to be 1. If it floated just half in and half out of water, we should know its specific gravity to be 0.5. The cylinder that we have to use will not float all in water or exactly half in water, but if we float it, and find the length of the part then in the water, we shall, by comparing this with the length of the whole cylinder, find some way of ascertaining the specific gravity of the cylinder.

Measure the length of the whole cylinder.

Float the cylinder in the jar (Fig. 18), keeping it upright by means of the holder, which is attached to the side of the jar. Joggle the cylinder to make sure that it is free to take its proper position. After each joggling it should come to rest at the same depth as before. The rings of the holder must not *grip* the cylinder at all. When sure that the cylinder is floating as it should, measure the length of the submerged part, from the bottom of the cylinder up to the *flat* surface of the water, not to the top of the *curve* where the surface meets the glass cylinder wall.

To find the specific gravity from the two measurements now made, begin by recalling the fact (see Exercise 8) that the water displaced by the floating cylinder weighs just as much as the cylinder itself.

How many times is the length of the submerged part of the cylinder contained in the whole length?

How many times the weight of the cylinder would be tne weight of a like cylinder of water?

What, then, is the specific gravity of the wooden cylinder?

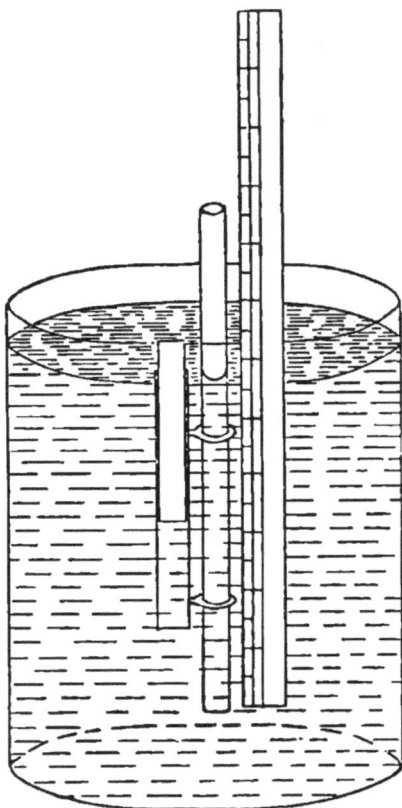

Fig. 18.

Suggestions for the Lecture-room.

PROBLEMS.

(1) A block whose specific gravity is 0.6 floats in water. How much of it is below the surface?

(2) A block whose volume is 1000 cu. cm., and whose specific gravity is 0.4, floats in water. How many cu. cm. of the block are below the surface?

(3) A block that weighs 4 oz. in air is fastened to a

sinker that weighs 6 oz. in water, and the two together weigh 3 oz. in water. What is the specific gravity of the block ?

(4) A block whose specific gravity is 0.5, and which weighs 100 gm. alone in air, is fastened to a sinker that weighs 150 gm. alone in water. How much will both together weigh in water ?

(5) A certain body has the density 187.2 in pound and foot units. What is its specific gravity ?

EXPERIMENTS.

Take two glass tubes, each about 6 in. long, connected by a rubber tube about 1 ft. long. Fill the whole with water, then close each end with a finger. Hold one end beneath the surface of the water in the gallon jar (Fig. 19); remove the finger from that end, and bring the other end, still closed, down outside the jar to a level lower than the water surface.

Is the water pressure against the finger that closes the tube now greater or less than the atmospheric pressure upon an equally l a r g e surface ? If greater, the water will run out when the finger is removed. If

FIG. 19.

less, the air will run in and drive the water up in the tube when the finger is removed. Try the experiment.

Repeat the experiment, but now hold the outer end of the tube, before opening it, higher than the level of the water in the jar.

A device like this, which is found in a great variety of forms, is called a *siphon.*

Show in operation glass models of "lifting-pump" (App. No. IX., Fig. 20) and force-pump (App. No. X., Fig. 21), discussing their action.

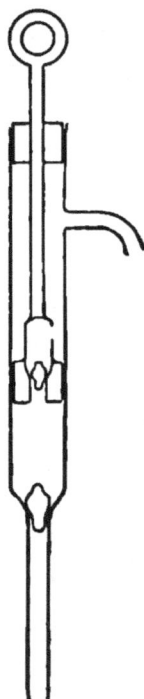

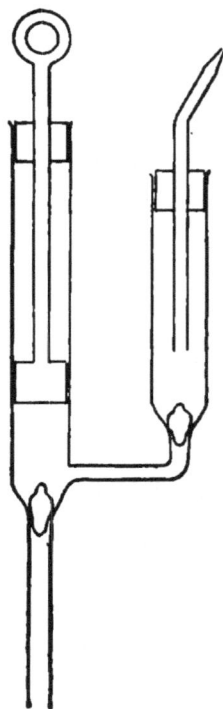

Fig. 20.   Fig. 21.

EXERCISE 12.

SPECIFIC GRAVITY OF A LIQUID: TWO METHODS.

Apparatus: The gallon jar (No. 10) nearly filled with water, and the smaller jar (No. 15) nearly filled with a solution of sulphate of copper.* The small glass bottle (No. 16). The spring-balance (No. 7). Thread.

---

* This solution may be made by putting 2 lbs. of sulphate of copper crystals into about 3 qts. of *warm* water in a glass vessel and stirring occasionally till the crystals are dissolved.

Weigh the bottle empty. Dip the bottle into the jar of sulphate of copper and let it fill with the liquid. Holding the bottle over the jar, put the stopper in place, thus crowding out the excess of liquid, then wipe the outside of the bottle and weigh it carefully with its contents.

Pour the sulphate of copper back into its jar, then fill the bottle with water, just as it was before filled with the other liquid, and again weigh the bottle and its contents.

From the three weighings now made the specific gravity of sulphate of copper can easily be found.

SECOND METHOD.

We found in Exercise 7 that a body going from air into water lost in apparent weight an amount equal to the weight of its own bulk of water. So a body going from air into a solution of sulphate of copper will lose in apparent weight an amount equal to the weight of its own bulk of the solution. This gives a method of finding the specific gravity of the solution. As a *body* to be weighed first in air, then in water, then in the solution, we will use the bottle with enough water in it to make it sink in either liquid. We may, indeed, use the bottle *full* of water, just as it was left at the end of the first part of this Exercise.

Suggestions for the Lecture-room.

PROBLEMS.

1. A glass sphere which weighs 100 gm. in air weighs 60 gm. in water and 40 gm. in sulphuric acid of a certain strength. What is the specific gravity of the glass?

What is the specific gravity of the sulphuric acid?

2. A vessel contains a layer of water 10 cm. deep and above this a layer of kerosene (sp. gr. 0.8), 10 cm. deep. What is the weight of a cube, each edge of which is 10 cm. long, that, if placed in this vessel, will sink till one half its volume is in the water and one half in the kerosene? *Ans.* 900 gm. What is its specific gravity? *Ans.* 0.9.

EXPERIMENTS.

Exhibit and show in operation two graduated glass hydrometers—one for determining the specific gravity of liquids less dense than water (App. No. XI.), the other for use with liquids more dense than water (App. No. XII.).

Show in a bottle together several liquids of different specific gravities that do not tend to mix with each other; for instance, mercury, chloroform, water, and kerosene.

Take a small tumbler containing some mercury and drop into it a piece of iron.

Take a bent glass tube (App. No. XIII., Fig. 22) each arm of which is about one foot long and pour water into it

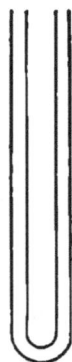

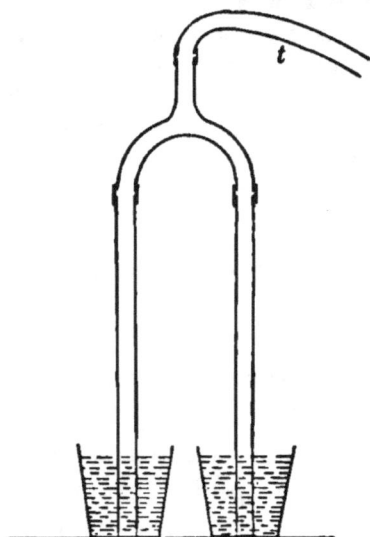

Fig. 22.   Fig. 23.

till both arms are about half full, then pour kerosene into one arm till it is nearly full. Does the water now stand as high in the other arm as the kerosene does in the first arm? Can you from this experiment see a third method for finding the specific gravity of a liquid?

Take a lead Y-tube and connect with two of the branches, by means of short rubber tubes, straight glass tubes about one foot long (App. No. XIV., Fig. 23). Attach to the third branch of the Y-tube a somewhat longer rubber tube, and let one glass tube stand in a vessel of water, the other in a vessel of kerosene or sulphate of copper. Apply the lips to the rubber tube *t* and draw out some of the air, taking care not to draw any liquid into the mouth. Note the height to which each liquid rises. Does this experiment suggest a method of finding the specific gravity of liquids?

CHAPTER III.

THE LEVER.

CIVILIZED men do most of their work with tools or machines. Many tools and many parts of machines consist of a piece of iron or wood or other material movable to a certain extent upon a support called a pivot, or axis, or fulcrum, by means of which a force applied in one direction at a certain spot may produce another force different in direction or in magnitude, or in both, at another spot. Such a tool or part of a machine is called a *lever*. One of the most familiar examples is a crowbar. A hammer, as used to *draw out* a nail from a board, is another example. Each half of a pair of scissors is a lever. We shall study some very simple forms of the lever to find out what relations hold between the forces exerted at different points.

EXERCISE 13.

THE STRAIGHT LEVER.

Apparatus: The lever and supporting bar (No. 17) fastened to the long horizontal bar that reaches above the table from end to end. Two scale-pans (Nos. 18A and 18B). A set of weights (No. 19).

Hang one scale-pan carrying a load of 8 oz. on the right-hand end of the lever at a distance of 14 cm. from the middle, as in Fig. 24.

Hang the other pan, with an equal load, on the left-hand end of the lever, at such a distance from the middle that the lever will *balance*, that is, stay horizontal when once placed so, even when the apparatus is jarred somewhat by tapping the short bar to which the lever is attached. Then make a record like this:

Left wt.	Left dist. fr. centre.	Right wt.	Right dist. fr. centre.
(1 + 8) = 9 oz.	. . . . .	(1 + 8) = 9 oz.	14.0 cm.

The space left blank here is to be filled by the left-hand distance which the pupil finds necessary to make the apparatus balance.

Change the right-hand weight to 7 oz., keeping its *place* unchanged, and move the left-hand weight, still 9 oz., to some new position which will make the whole balance, in spite of jarring, as before. Make a record, as before, of the weights and distances, putting it just beneath the record for the first arrangement.

Change the right-hand weight to 5 oz. without changing its place and find what position the left-hand weight, still remaining 9 oz., must have in order that the lever may balance. Record the

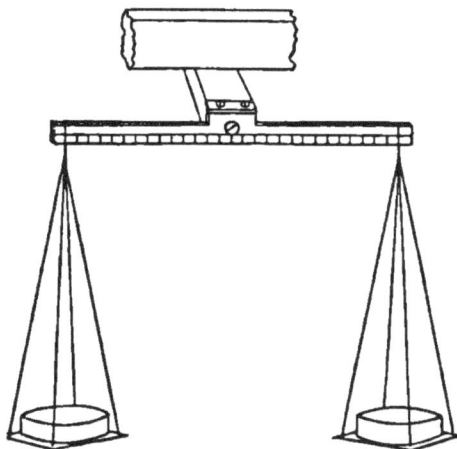

Fig. 24.

distances and weights for this case under the records already made for the first and second cases.

One more case may be taken in which the right-hand weight becomes 3 oz., still at 14 cm., which will give a fourth line in the record table. More observations with different arrangements might be made, but it is better to make a moderate number of good observations than a large number of hasty or careless ones.

By studying the record table now made the pupil may find a rule by which, when the two weights and one distance are given, the other distance may be found by calculation; or when the two distances and one weight are given the other weight may be found by calculation.

Suggestions for the Lecture-room.

In the preceding Exercise the class found out how to make the two weights hung from the lever balance each other. Let us ask now what the rule for balancing would be, if there were more than two weights in use, as in Fig.

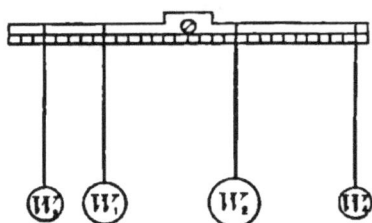

FIG. 25.

25, for instance. We will make the apparatus balance with four weights, two on each side. We will call the weight nearest the centre on the left hand weight No. 1, which we will write W_1, for short. The other weight on the left-hand side we will call No. 3, or W_3. The two weights on the right hand we will call W_2 and W_4.

When the whole balances, we will call

the distance of W_1 from the middle D_1,
"　　"　　" W_2　"　　"　　" D_2,
"　　"　　" W_3　"　　"　　" D_3,
"　　"　　" W_4　"　　"　　" D_4.

Now if we go back for a moment to the case of two weights, which the class has studied, and if we call these now P_1 and P_2, and their distances from the middle d_1 and d_2, we can state the rule for balancing in this way :

$$P_1 \times d_1 \text{ must equal } P_2 \times d_2.$$

In the new case, where we have four weights, we may *guess* that the rule is

$$(W_1 \times D_1) + (W_3 \times D_3) = (W_2 \times D_2) + (W_4 \times D_4),$$

and then test the truth of our guess by trial.

Other cases may be tried until a satisfactory conclusion can be arrived at.

In the experiments with which we have just been engaged the weights have been suspended from the top of the lever on a level with that part of the pivot upon which the lever rests. In other experiments which are to follow we shall not always be able to keep this arrangement, and we have now to find out what would be the effect of hanging one or more of the weights from points higher or lower than the point of support of the lever. For this purpose we shall use No. XV., the piece of apparatus shown in Fig. 26, in which the straight lever thus far used is replaced by a circle of wood about 8 inches in diameter, supported by a screw passing horizontally through the centre. Such a circle, or disk, of wood comes under the general definition of a lever.

We will hang at b and f such weights as will balance each other, leaving the disk in equilibrium, and will then move one

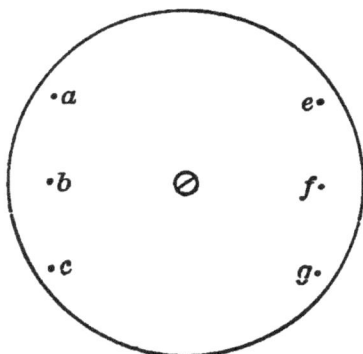

FIG. 26.

of the weights to a point vertically above or vertically below its present place; that is, from f to e or g, or from b to a or c. Shall we still have equilibrium?

We will now turn the disk a little, so that the lines $a\,b\,c$ and $e\,f\,g$ will be no longer quite vertical, and will see whether now a weight at e or at g has just the same effect as if at f.

A careful note should be made of the conclusions arrived at, for future use.

In the experiments upon the lever thus far, the lever itself, whether a bar or a disk, has *balanced*, when left to itself without load. We have, therefore, not had to consider the weight of the lever itself. But many levers are used in such a way that their own weight helps or hinders the operation to be performed with them. To understand such cases we must learn something about what is called the *centre of gravity* of a body.

EXPERIMENT.

Take a board, cut in any irregular shape, like Fig. 27, for instance. Make several small holes *straight* through

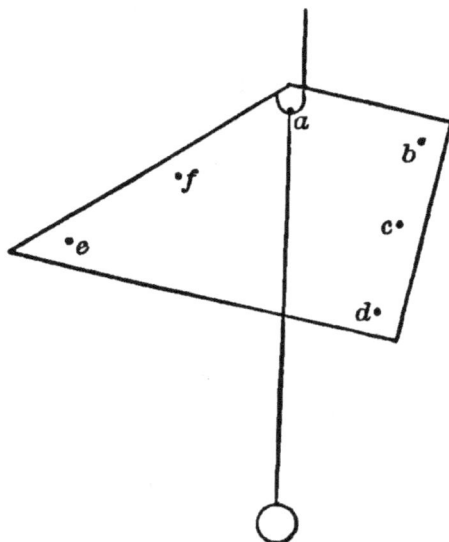

Fig. 27.

the board, and put into each hole a wire nail that will fit close, long enough to project about half an inch on each side of the board. Tie a bullet at one end of a

Fig. 28.

thread and make a loop in the other end. Put this loop over one hook of a piece of wire bent into the shape shown in Fig. 28, and then rest the nail *a*, Fig. 27, in the hooks of the same wire, so that the board and the string carrying the bullet will both hang free, the string near the face of the board. (The whole apparatus as shown in Fig. 27 will be called No. XVI.) Mark with a pencil the course of the string downward across the board.

Then suspend the board by the nail *b* and mark the new course of the string. Proceed in this way with all the nails and note the point where the various pencil-marks cross each other.

Finally, place the board horizontal and balance it upon the flat head of a lead-pencil, noting how near the head of the pencil comes to the crossing of the lines marked on the board.

By such experiments as this we come to see that there is within the board a certain point which always hung just beneath the support when the board came to rest suspended from any one of the nails. We see that the same point has to be just *above* the support when the board rests upon the pencil-top. In short, the board *acts* in these experiments as it would if all its weight were concentrated at this particular point. This point might be called the centre of weight or centre of heaviness of the board, but it is commonly called the *centre of gravity.*

The following exercise is intended to make the pupil more familiar with the idea of centre of gravity, and to show how it may be taken account of in the use of the lever.

EXERCISE 14.

CENTRE OF GRAVITY AND WEIGHT OF A LEVER.

Apparatus : The lever of No. 17, detached from its supporting bar, and a small block (No. 20), the two being fastened together, as in Fig. 29, so as to make one body, which will be called the

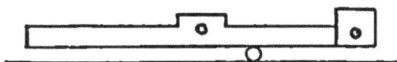

Fig. 29.

lever in this exercise. A slender wooden cylinder (No. 13). A 1-oz. scale-pan (18A or 18B). A 1-oz. wt. from No. 19,

To find the centre of gravity of the lever, balance it, bar and block fastened together, as nearly as you can in a horizontal position on the cylinder laid on the table (see Fig. 29), the cylinder being kept at right angles with the lever. Find in this way at what particular mark of the bar the centre of gravity is, and record this mark—for instance, 9.1 cm.

Then suspend the 1-oz. scale-pan carrying a 1-oz. wt., two oz. in all, from any convenient point near the free end of the bar, and letting this end project beyond the edge of the table-top, balance the whole, as now arranged, as nearly as you can, on the cylinder laid on the table as before (see Fig. 30).

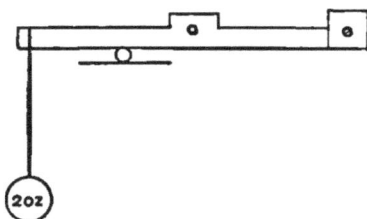

FIG. 30.

Now record the mark from which the scale-pan hangs, 33.4 cm., we may suppose, and the mark which is just over the middle of the cylinder when the whole balances, 21.6 cm., let us say.

This case is like that of the lever studied in Exercise 13. The cylinder now taking the place of the screw as a support, we see that

the left-hand weight is 2 oz.,
"   "   "   distance is 33.4 − 21.6 = 11.8 cm.,
" right-hand weight is the weight of the lever,
"   "   "   distance is 21.6 − 9.1 = 12.5 cm.,

that is, the distance from the support in Fig. 30 to the centre of gravity of the bar and block.

We do not as yet know the weight of the lever, but we will call it W_2, and see whether we can find its amount by calculation. If we apply the same rule that was found to hold true in Exercise 13, we shall have

$$2 \times 11.8 = W_2 \times 12.5,$$

which gives for the weight of the bar and block

$$W_2 = \frac{2 \times 11.8}{12.5} = 1.89 \text{ oz., nearly.}$$

The value of W_2 obtained in this way by the pupil should be compared with the weight of the bar and block as found by the teacher with some balance, e.g. No. **XVII.**, much more sensitive than the spring-balance used by the class, for if the method of this Exercise is carefully followed it will give the weight of the lever more accurately than the spring-balance is likely to do.

Suggestions for the Lecture-room.

In connection with the idea of centre of gravity discuss and illustrate *stable equilibrium, unstable equilibrium,* and *neutral equilibrium* (see any ordinary descriptive text-book of Physics).

We have now found out how to take account of the weight of the lever itself, when we need to do so. We know that all its weight may be regarded as concentrated at a certain point, which we call the centre of gravity, and we have tried one case in which the weight of the lever itself, acting at the centre of gravity, balanced a certain weight suspended from the bar. In common levers, like the crowbar, the weight of the bar itself may be very important sometimes, when the fulcrum is a long distance from the centre of gravity of the bar.

We will now return for the present to cases where the centre of gravity lies, as in Exercise 13, just under the point of support of the bar. In this case the weight of the lever itself does not tend to make the bar tip in either direction from its horizontal position.

CLASSES OF LEVERS.

In the levers which we have studied thus far the support, or *fulcrum* as it is called, lies between the lines of suspension of the two weights. This kind of lever, whether it is a simple bar or a disk or an object of irregular shape, whether its centre of gravity lies at the point of support or not, is called a lever of the *First Class*.

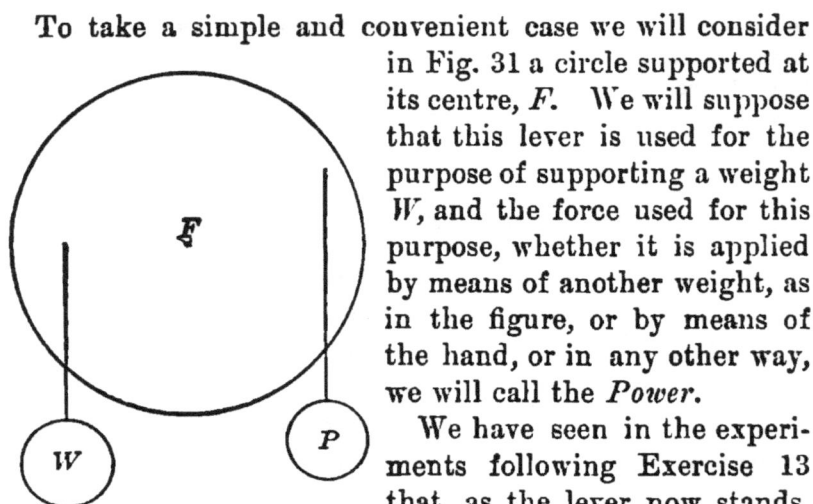

To take a simple and convenient case we will consider
in Fig. 31 a circle supported at
its centre, *F*. We will suppose
that this lever is used for the
purpose of supporting a weight
W, and the force used for this
purpose, whether it is applied
by means of another weight, as
in the figure, or by means of
the hand, or in any other way,
we will call the *Power*.

We have seen in the experiments following Exercise 13
that, as the lever now stands,
it makes no difference whether

FIG. 31.

W is suspended from the point which now carries it or
from some point higher or lower in the same vertical line,
which is called the *line of action* of *W*. A like statement
can be made for *P*. We shall call the shortest distance
from *P*'s line of action to the fulcrum the *power-arm*, and
the shortest distance from *W*'s line of action to the fulcrum
the *weight-arm*.

In order that *P* and *W* may just balance each other we
must have, as can be seen from Exercise 13,

$$power \times power\text{-}arm = weight \times weight\text{-}arm.$$

This is the *law* for a lever of the *First Class*.

But we may have a case, like that shown in Fig. 32, in
which the line of action of the weight lies between the fulcrum and the line of action of the power. This arrangement gives us what is called a lever of the *Second Class*.

There is still a different case, shown in Fig. 33, where the
line of action of the power lies between the fulcrum
and the line of action of the weight. This is called a lever
of the *Third Class*.

In the second and third classes of levers, as in the first class, the shortest distance from the fulcrum to the line of action of the *weight* is called the *weight-arm*, and the

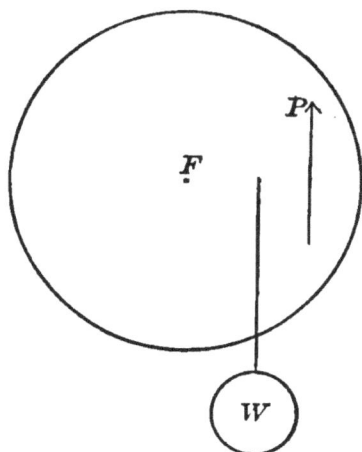

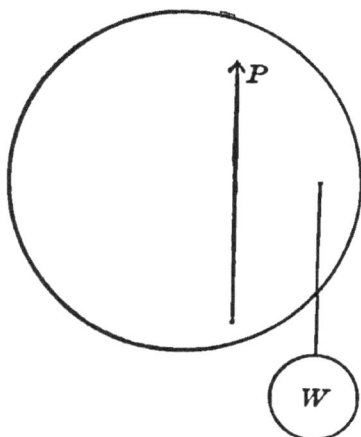

Fig. 32.   Fig. 33.

shortest distance from the fulcrum to the line of action of the *power* is called the *power-arm*.

The pupil is to find out by means of the following Exercise whether the laws of the second and third classes of levers are as simple as the law of the first class.

EXERCISE 15.

LEVERS OF THE SECOND AND THIRD CLASSES.

Apparatus : The lever (No. 17) supported as in Exercise 13. A scale-pan (No. 18). A set of weights (No. 19). A spring-balance (No. 7).

Suspend the pan with a load of 8 oz. at a point 5 cm. from the middle of the lever, and, on the same arm of the lever, at a distance of 10 cm. from the middle, pull upward with a spring-balance, connected with the lever by means of a loop of thread, until the weight is balanced and the lever becomes horizontal. You have here a lever of the second class. Read the spring-balance and record as follows :

LEVER OF SECOND CLASS.

Weight.	Weight-arm.	Power.	Power-arm.
9 oz.	5 cm.	....	10 cm.

Try other similar cases, and study them all until you are able to write down the law for this class of levers.

Then with the same apparatus place the spring-balance between the fulcrum and the line of the weight. You will now have a lever of the third class. Try various cases and record as before.

LEVER OF THIRD CLASS.

Weight.	Weight-arm.	Power.	Power-arm.
....	....	....	....
....	....	....	....

Law..

Suggestions for the Lecture-room.

PROBLEMS.

1. A lever supported at its centre of gravity is used to lift a weight of 100 lbs. applied at a distance of 1 ft. from the fulcrum. The power is applied 5 ft. from the fulcrum and on the opposite side from the weight. How great must the power be? Must the power be applied upward or downward?

2. If the power had been placed on the same side of the fulcrum as the weight, everything else being as described in the preceding problem, how great would the power have to be? Would it be applied upward or downward?

3. If the power were 50 lbs. applied 2 ft. from the fulcrum toward the right, and if the weight were applied 8 ft. from the fulcrum toward the right, how great could the weight be?

4. If a weight of 5 lbs. were placed 4 ft. toward the right from the fulcrum, and a weight of 7 lbs. 6 ft. toward the right from the fulcrum, how far from the fulcrum toward the left must a force of 10 lbs. be applied in order to make the whole balance?   *Ans.* 6.2 ft.

In the four preceding problems the weight of the lever has not been considered, because the centre of gravity has been supposed to be at the point of support. Suppose now that the lever weighs 4 lbs. and that its centre of gravity is 3 ft. to the right from the fulcrum, and with this new condition go over each of the four problems again.

If we take a case like that shown in Fig. 34, it is plain that 4 oz. applied 7 cm. from the centre will balance 2 oz. applied 14 cm. from the centre, but it may not be perfectly plain how great the pull on the fulcrum is in this case. We will, therefore, in the next Exercise try the experiment in one or two simple cases and see what the result will be.

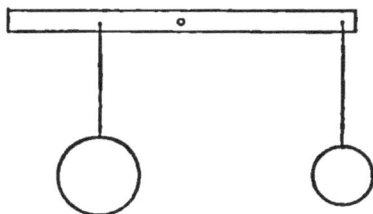

FIG. 34.

EXERCISE 16.

FORCE EXERTED AT THE FULCRUM OF A LEVER.

Apparatus: The lever of No. 17 freed from its support. Two scale-pans (Nos. 18A and 18B). Two 1-oz. wts. and one 2-oz. wt. from No. 19. The spring-balance (No. 7). A piece of copper wire about 1 mm. in diameter bent into the form of a hook (h in Fig. 35). A piece of thread about 6 inches long.

Suspend the bar from the balance in the manner shown by Fig. 35. Note and record the weight of the bar alone. Then suspend one scale-pan with a 1-oz. weight from one arm of the bar, and the other scale-pan with a 2-oz. weight from the other arm in such a way as to balance, taking care not to let the pans and weights spill. Note and record the reading of the balance. Then make the loads (pan and weight) 2 oz. on one side and 4 oz. on the other, and read

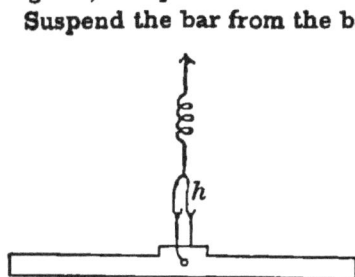

FIG. 35.

and record. **Try any other experiments that you can with the weights furnished, until you feel reasonably sure that you know the relation between the weights applied and the pull on the balance. Then state what this relation is.**

Suggestions for the Lecture-room.

In each of the cases in Exercise 16 we have applied two downward forces to the bar in suspending the two scale-pans with their loads, and have found these two forces to be balanced by another force exerted upward by the spring-balance. It will be well for us to study such cases very carefully, for similar ones are often found.

Suppose we are to make three parallel forces, A, B and C, just balance each other when all are applied to the same body. Can we from what we have now learned tell anything about the relative magnitude and the arrangement of these forces?

We know that—

1*st*. *All the forces cannot point in the same direction.* Let us suppose that C is opposite in direction to A and B.

2*d*. *The force C must be equal to the sum of the two forces A and B.*

3*d*. *The line along which C is applied must lie between the lines along which A and B are applied.*

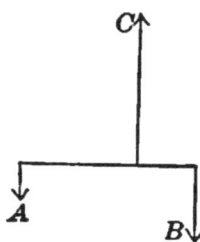

Fig. 36.   4*th*. $A \times$ *shortest distance from line of A to line of C* $= B \times$ *shortest distance from line of B to line of C.* (See Fig. 36.)

These rules apply as well to horizontal forces as to vertical forces. Try three spring-balances laid parallel to each other on a table and pulling at some light horizontal bar—a lead-pencil, for instance.

We have already learned to consider a disk pivoted at

the centre as a kind of lever. When such a lever is worked by means of strings lying upon its circumference it is called a pulley. We shall now see that the pulley form of lever has some great advantages.

Take the pulley shown in Fig. 37 (No. XV), and let us

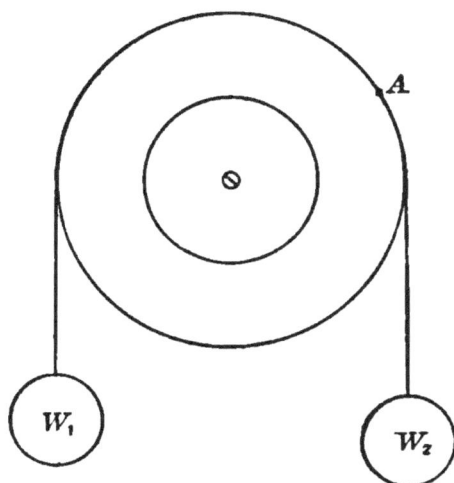

FIG. 37.

first use the largest circle only. If we fasten two equal weights, W_1 and W_2, to the ends of a string and pass the string across the top of the pulley, we shall of course find that they balance each other. But suppose we used two strings, one for W_1 and the other for W_2, fastening each string to a pin or tack at point A, but letting each string rest in the groove of the pulley, so that the final position of the two strings will be represented by Fig. 37. Will two equal weights balance each other under these conditions? The question is quickly answered by trial, and by turning the pulley a little one way or the other one may try the experiment with A in a variety of positions.

Next try the effect of a horizontal pull, *P*, applied by a spring-balance at the top of the pulley to balance a weight,

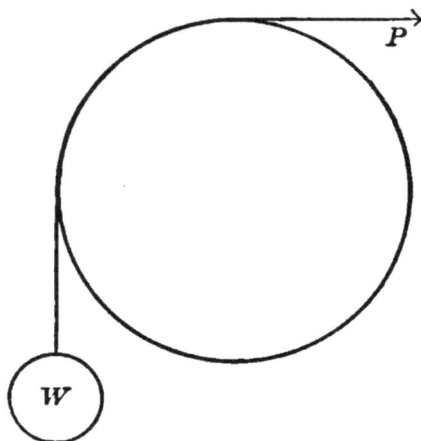

W, see Fig. 38. (Remember that in this position the *read-ing* of the *ordinary* 8-oz. balance is about ½ oz. less than the real force exerted by it, because the spring of the balance does not now support the weight of the hook and bar, which is about ½ oz.). Find by experiment whether the force *P* must be greater or less than, or equal to, the direct pull of the weight *W*.

Various experiments interesting to a class can be made by balancing a weight on one circle of the pulley by a weight on another circle, and the simple rule which holds for the relation between the balancing weights is easily made out.

We see that the advantage of a pulley like this, as com-pared with a simple bar lever, is that the pulley enables us to vary the direction of our power at will and to *lift* a weight a much greater distance than we could with a bar lever no longer than the diameter of the pulley. In fact, the distance through which we can lift the weight by means of the pulley depends merely upon the length of the string that supports the weight.

But we can do more with a pulley than we have yet done. Let us take now a well-made small metal pulley (No. XVIII), such as one may get at a hardware store, and arrange it according to the indications of Fig. 39, *P* being the pulley, *M* a weight suspended from an axis through the centre of the pulley, *B* a spring-balance, and *h* a hook to which one end of the string passing beneath the pulley is attached.

To what class of levers does the pulley in this position belong? What, then, should be the relation between the weight, which is *M plus* the weight of the pulley itself, and the pull exerted by the spring-balance? Find by experiment whether the conclusion reached is correct. (In making this trial one must remember that friction is often large in cheap pulleys, even when they are well oiled, as this one should be. Now when the load is being steadily raised, the hand carrying the spring-balance must lift harder than it would if there were no friction, but when the load is being steadily lowered, the hand, pulling just hard enough to prevent the load from *hurrying*, is assisted by the friction. The mean between the reading of the balance going up and the reading of the balance coming down will show, very nearly, what the pull required to sustain the load would be if there were no friction.)

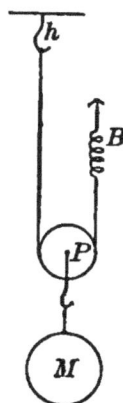

Let us now try an arrangement like that shown in

FIG. 39.

Fig. 40, in which we have one pulley, *A*, hooked to a bar overhead and a double pulley, *B*, which moves up and down with the load (No. XIX) *M*. Let us consider what should be the relation between the pull *P* and the weight *W*, which is *M plus* the weight of *B*, in this case.

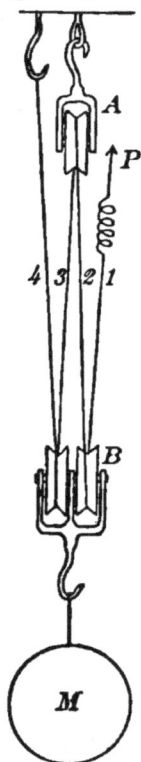

In the case tried just before we had two strings holding up the pulley *P*. We have now four strings holding up the pulley *B*. After thinking upon the matter for a little time, trying to study out what is the relation between *P* and *W* with this arrangement, let us try the experiment as we have already tried it in the simple case, noting the force shown by the spring-balance when *M* is moving steadily up and again when it is moving steadily down, and taking the mean between these two forces as the one that would be required to balance the weight, *W*, if there were no friction.

FIG. 40.

Can the class name any tools or machines, not already mentioned in this book, in which levers or pulleys are used?

CHAPTER IV.

THREE FORCES WORKING THROUGH ONE POINT.

In studying the lever we have usually, though not always, had parallel forces to deal with, forces acting straight up or straight down. But very often we have to do with bodies that are acted upon by forces not parallel to each other. Thus when a ladder standing upon the ground leans against a house, we have at least three forces acting upon the ladder: 1st, the earth's attraction, or, as we call it often, the *weight* of the body, which acts as if the whole substance were at the centre of gravity ; 2d, the push of the ground against the foot of the ladder, which push is not straight upward ; 3d, the push of the wall against the top of the ladder.

Again, a flying kite is acted upon by the earth's pull, straight downward ; by the force exerted by the air, which force, because of the wind, is not straight upward ; by the pull of the string, which pull is not straight downward.

The way to begin the study of such cases is to study the case of three forces all acting straight from or straight toward a single point. We shall take such a case in Exercise 17.

EXERCISE 17.

THREE FORCES IN ONE PLANE AND ALL APPLIED TO ONE POINT.

Apparatus : Three 8-oz. spring-balances, each provided with two small blocks (No. 21) to go under its sides and hold it flat on its back when it is lying upon the table. The rectangular block (No. 9). The measuring-stick (No. 3). A sheet of paper. Thread,

Take two pieces of strong thread, one about 12 inches, the other about 6 inches, long, and tie one end of the short thread to the middle of the long one. Fasten the three loose ends to the hooks of the spring-balances, then lay the latter upon the table, putting the blocks under their sides, as in Fig. 41, and let one student pull at each balance, taking care that the slit of each balance-face is in a straight line with the thread, until no one of these reads less than 3 oz. It will be found that any variation in the angles

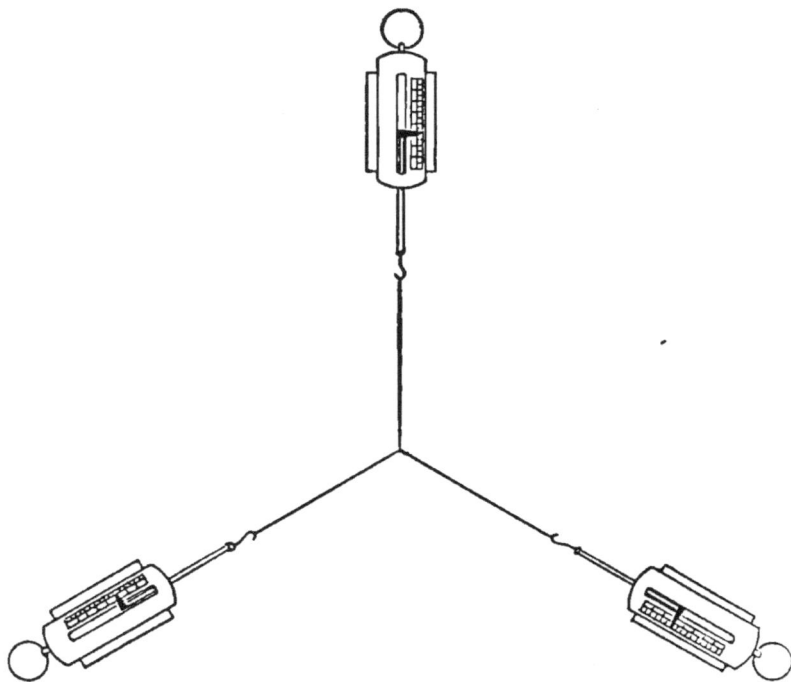

8-oz. Spring-balances.

Fig. 41.

which the strings make with each other will require a change in the forces. Evidently there is some connection between the directions of the strings and the forces necessary to balance each other. The object of this exercise is to make out what this connection is. It is simple and easy to remember. We can study it best if we put under the threads a sheet of paper and draw on this paper, just under each thread, a pencil-mark parallel to the thread, and then write down alongside each pencil-mark the force in the direction of that line, as shown by the spring-balance. The balances must be held very still while these lines are being drawn and must be read before any change occurs in the direction of the lines. (It may prove best to fasten the *ring* of each balance to a weight heavy enough to hold the balance in place, thus relieving the pupils, who might grow tired and unsteady in holding the balances long enough to permit of drawing the pencil-marks properly.) To draw a line, place one side of the block (No. 9) close alongside one branch of the thread, taking care not to push the thread out of place, and then run the point of a well-sharpened pencil along the edge of the block under the thread. Draw the other lines in the same way, doing it all very carefully.

Each pupil in turn makes a set of lines, and records alongside

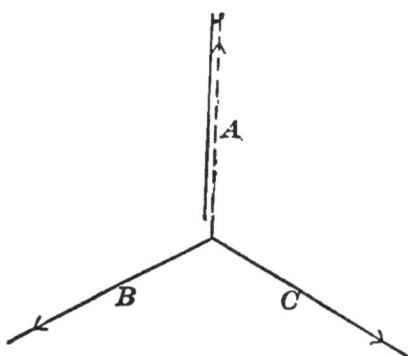

FIG. 42.

them the proper forces. The directions of the pulls should be varied somewhat by each pupil, in order that his lines and forces may not be exactly like those of other pupils.

Take now the wooden ruler (No. 3), and extend the three lines toward each other till they meet at one point. This they will do if they have been drawn originally just *under* the threads. If they do not all meet in one point, a new line should be drawn parallel to one of them, which new line will pass through the crossing of the other two lines, and this new line, the dotted line in Fig. 42, is then to be used in place of the original line. The three lines as now drawn will represent accurately the directions of the three forces.

Now measure off from the common point along the line *A* a distance of 1 cm. for each ounce (or each 30 gm., if the forces are measured in grams) of the force which was exerted along that line, and put a small arrow-head (see Fig. 42) at the end of this measured distance. Erase that part of line A which lies beyond the arrow-head.

Do the same with lines B and C that has been done with A. The three arrows thus obtained, all reaching from the same point, represent the magnitude and the direction of the three forces exerted by the spring-balances.

Now with A and B of Fig. 42 as two of the sides make a parallelogram, taking pains to make it accurate.* Then make a parallelogram with B and C as sides; then one with A and C as sides. Compare the length and direction of the line C with the length and direction of the diagonal of the parallelogram AB; the line A with the diagonal of the parallelogram BC; the line B with the diagonal of AC.

---

* One line may be drawn very nearly parallel to another by means of a device illustrated by Fig. 43. Ll is a line already drawn. The

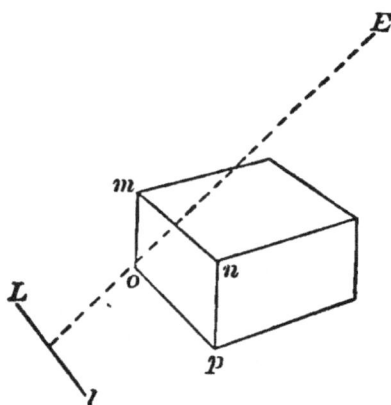

Fig. 43.

block (No. 9) is so placed that for an eye placed at E the edge mn appears to be close to Ll and parallel to it. Then a pencil-mark is made along the edge op.

A better method is to set the edge op on the line Ll and then guide the block to a new position by sliding it along the straight edge of a ruler.

From this comparison make a rule showing how to find the direction and magnitude of a force *C* which, put with two forces represented by the lines *A* and *B* (Fig. 44), will just balance them.

Suggestions for the Lecture-room.

Problem for the class, to be solved by simple calculation or by drawing a figure and measuring: A force of 7 lbs. pulls north from a certain point and a force of 4 lbs. pulls east from the same point. How large must a third force be to hold them in check, and what will be its general direction?

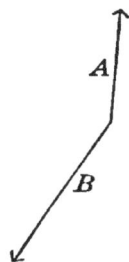

FIG. 44.

THE INCLINED PLANE.

The facts learned in Exercise 17 will enable us to understand a contrivance very often used for raising heavy weights. You have all seen barrels of flour or other heavy objects loaded upon wagons by rolling them up a plank or a pair of rails, placed with one end on the ground and the other upon the wagon, so as to make the ascent gradual instead of straight up. The flat slanting surface up which the body is rolled is called an *Inclined Plane.*

Sometimes a body is lifted by forcing an inclined plane, the slanting face of a *wedge,* under it, as in Fig. 45.

FIG. 45.

Sometimes the force used by an experimenter or a workman with the inclined plane is parallel to the inclined surface; sometimes it is parallel to the base-line of the plane, the horizontal surface of a wedge, for instance, in Fig. 45,

It is well known that the force required to move a body
up an incline, or to keep it from sliding down the incline,
is greater the greater steepness of the incline. The experiments now to be undertaken are for the purpose of
making out the connection between the weight, the steepness of the incline, and the power required to hold the
weight from sliding or rolling down the incline, when
there is no friction to oppose this motion.

We shall consider first the case in which the power is
applied parallel to the inclined surface.

Take apparatus No. XX and adjust it as indicated by
Fig. 46, putting 7 oz. upon the pan, so that $P = 7 + 1 =$
8 oz. Then raise or lower the incline till the weight W

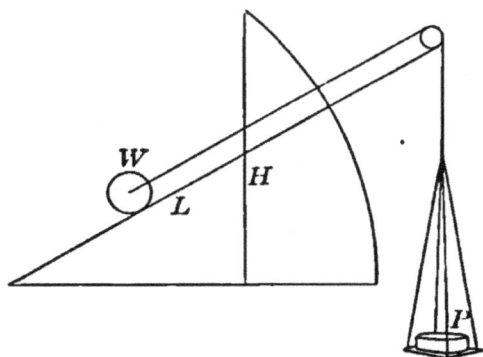

Fig. 46.

will barely roll up the incline when the apparatus is pur-
posely jarred slightly. (The incline cannot be quite so
steep when this takes place as it might be if there were no
friction. If a knot is made in the thread near where it
passes over the pulley at the top of the incline, a very
slight movement up or down the incline can be detected
by watching the position of this knot. A slight movement
is enough. It is not necessary to have the weight W move
far.)

As soon as this adjustment is made, read *H*, the length of the vertical scale from the top of the base-board to the under side of the incline, and record in the way indicated in the table below (upper row of numbers).

Then without changing *P* raise the incline somewhat more, until *W* will, when the apparatus is jarred, barely roll down the incline. (The incline must be somewhat steeper for this than it would have to be if there were no friction.) When the proper adjustment is made read the new value of *H* and record it in the second line of the table below.

To find the *H* that would make *P* just balance *W*, if there were no friction, take the mean between the two values now recorded. Then find the *L* that would correspond to this value of *H*, *L* being the distance along the inclined scale from the hinge to the point of crossing the vertical scale.

	P	*W*	*H*	*L*
Going up.......	8 oz.	16 oz	....	
" down....	8 "	16 "	....	
To balance	8 oz.	16 oz.	.... (mean)	....

Then make *P* = 6 oz., then 4 oz., and in each case repeat the operations just described.

Then bring together the results of all the observations in the following form:

	P	*W*	*H*	*L*
Required to balance {	....	....	....	....
	....	....	....	....
	....	....	....	....

It will then, probably, be easy to make out the law which holds in this application of the inclined plane.

For experiments in which the power is applied parallel to the base-line we cannot make use of a string running over a pulley. We must apply the power by means of the spring-balance, as shown in Fig. 47, the long slot cut through the incline lengthwise allowing us to do so.

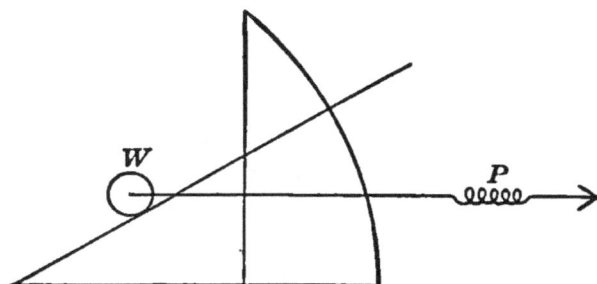

FIG. 47.

Find by trial a steepness of incline that will make P about 7 oz., and, keeping this steepness unchanged for the time, find how large P is when it is pulling W slowly and *steadily* up the incline, and how large when it is letting W run with equal slowness and steadiness down the incline. Take the mean of these two values as the one that would be needed to balance W if there were no friction. (The mean of the two values of P is not, in this case, exactly the quantity wanted, because the greater pull of P when W is going up the incline makes W press harder against the incline when going up than when going down. The mean value of P, as now found, is a little greater than the value wanted, but so little that the error is not important.)

We record, then, for this case:

	P	W	H	B
Going up.....	....	16	....	
" down ..	....	16	....	
To balance ...	.... (mean)	16	....	....

where B is the length of the *base-line* from the hinge to the foot of H.

Lower the incline and try various degrees of steepness, so that P will be in one case about 5 oz. and in another case about 3 oz. Then arrange the results of the various cases tried in this form:

	P	W	H	B
	....	....	....	....
Required to balance	....	....	....	....
	....	....	....	....

Look for the law here, and state it when found.

With the aid of a little knowledge of geometry the laws of the inclined plane might be found without these experiments, from the law of the parallelogram of forces. Fig. 48 suggests the reasoning for the case of a pull paral-

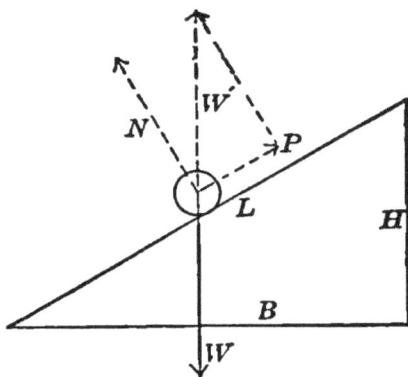

Fig. 48.

lel to the incline. W here represents the *weight* of the body exerted straight downward from the centre of gravity. The dotted line W' is equal to W, but opposite in direction. It is the diagonal of a rectangle having N and P as sides. N represents the force exerted upon the roller by the plane L, a force which is straight outward from the plane L, if there is no *friction* (see the next Exercise) of the roller against the plane. P represents the pull, parallel to the plane L, which with the force N will just balance W. Compare the dotted triangle with the triangle whose sides are L, B, and H, and see whether you can by use of your geometry make out the relation between P, W' ($= W$) and certain sides of the triangle LBH.

Fig. 49 suggests the reasoning for the case of a pull parallel to the base.

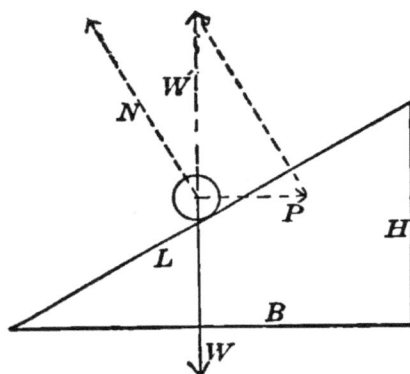

Fig. 49.

CHAPTER V.

FRICTION.

WHEN we push a heavy block along on the top of a table we feel a certain resistance. We know from experience that by making the surface of the table and the surface of the block very smooth we can lessen the resistance. This resistance, the amount of which depends upon the condition of the rubbing surfaces, is called *Friction.*

Friction always opposes motion, whatever may be the direction of the motion. That is, it merely tends to *stop* the motion. It never helps to push the block *back* to the position where it started.

We shall measure in a number of cases the force required to keep a block moving steadily along on a sheet of paper laid upon a level table-top.

EXERCISE 18.

FRICTION BETWEEN SOLID BODIES.

Apparatus: A spring-balance (No. 7). A rectangular block (No. 9). Set of weights (No. 19). A smooth sheet of paper about 1 ft. wide and 1½ ft. long. Thread.

We shall first consider the *velocity* of the motion. That is, we shall ask whether the force required to keep up a slow steady motion is greater or less than that required to keep up a more rapid steady motion.

Lay the block on one of its broad sides, and attach it to the spring-balance by a thread passing around but not *under* the block. Load the block with weights until the force required (for a very slow, steady motion) is about 3 oz. Draw the block parallel to its *grain* along the sheet of paper several times with a very slow, steady motion, and several times with an equally steady motion two or three times as fast. As the paper is likely to grow somewhat smoother under the repeated rubbing, the experimenter should not make all his slow trials first, but should change from slow to fast and fast to slow a number of times.

Record your conclusion as to whether the slow or the more rapid motion requires the greater force.

We shall next try to find out whether, the total weight being the same as before, it is easier or harder to draw the block on a narrow side than on a broad side. Use the same block and the same load of weights, pulling it now as before parallel to its grain.

Of course the side upon which the block slides should in all cases be clean, and the broad and narrow sides which are compared should be, as nearly as practicable, equally smooth. The thread must not be between the rubbing surfaces in any case.

Record your conclusion as to whether the broad side or the narrow side offers the greater resistance to the motion.

Finally, we shall ask what connection there is between the weight drawn and the force required to draw it. For this purpose vary the weights placed upon the block, using not less than 6 oz. for the least and as much as 16 oz. for the greatest load.

Add to the load in each case the weight of the block itself and make the record in the following form, W being the load and b the weight of the block:

$W + b.$	F (Force required).
. . . .	. . . . .
. . . .	. . . . .
. . . .	. . . . .

Look for any simple relation between $(W + b)$ and F.

The experiments just described will teach a number of useful facts about friction between two solid substances, but one must be careful not to apply the conclusions here arrived at to extreme cases, extremely slow or very fast motion, for instance; or to cases where the pressure is great enough and the edge of the sliding body narrow enough to cause an actual cutting of the body into the surface over which it should slide.

CHAPTER VI.

THE PENDULUM.

Before leaving the subject of Mechanics and going to that of Light it is well to learn something about pendulums, which are used to control the motion of clocks.

If you were to examine the works of an old-fashioned clock you would find the power which drives it in a heavy weight working upon a kind of pulley by means of a long cord, but the device which governs the speed of the works and allows the motion to be neither too fast nor too slow is the pendulum. As a crowd of men at a turnstile, however they may try to force their way, can pass no faster than the swinging turnstile permits, so the clock-weight, which if the control were removed would run down at once with a furious buzzing of the wheels, is allowed by the pendulum to descend only very slowly, a very little distance at every swing of the pendulum, and not at all when the pendulum does not move.

The rate at which the clock-wheels can move, then, depends upon the length of time required for each swing of the pendulum. We will try a few simple experiments with very simple pendulums to find out—

1st. How does the time required for a single swing depend upon the *length*, or *width*, of the swing?

For this purpose we have, hanging side by side, two pendulums of equal length, Nos. 1 and 2 in Fig. 50, each consisting of a bullet suspended by a silk thread about 3 feet long.

(A convenient method of suspending each pendulum is

shown in Fig. 51, where *B* is one end of a wooden bar,
which is bevelled off on the side from which the pendulum
hangs. *C* is a cork fastened to the top of the bar and having in it a slit made by a sharp knife, through which slit
the silk thread, *S*, passes. If this part of the thread is

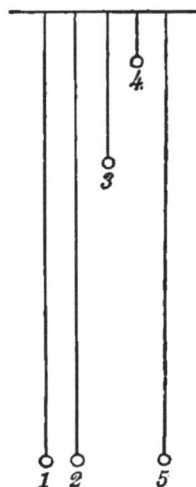

Fig. 50.   Fig. 51.

waxed, the fastening thus obtained holds the pendulum
securely, although it is very easy to increase or decrease the
length of the pendulum at will. The length of the pendulum is to be measured from the under side of the bar to
the centre of the ball. It is intended that the length of
No. 3 in Fig. 50 shall be one fourth that of No. 2, and the
length of No. 4 one ninth that of No. 2. It is, therefore,
convenient to make the length of No. 2 just 36 inches,
which will require 9 inches for the length of No. 3, and 4
inches for that of No. 4. The suspended body is a bullet
in the case of each pendulum except No. 5, where it is some
lighter object—a marble for instance. No. 5 has the same
length as No. 1.)

The five pendulum balls and the bar *B*, prepared for receiving the threads, will together be called No. XXI.

We will first set No. 1 and No. 2 swinging at the same instant and with the same width, or length, of swing, and watch them both for a little while until we see that under these circumstances they keep together, No. 1 taking just as long a time for one swing, or for any number of swings, as No. 2 does.

Then draw the ball of No. 1 about one inch aside from its position of rest, and the ball of No. 2 about fifteen inches aside from its position of rest, and release both balls at the same instant. The class will watch the two for some little time, a quarter of a minute or longer, and see whether at the end of that time they begin each swing together, as they did at first. If they do not, observe which one has gained upon the other, and, after one or two repetitions of the experiment, write down an answer to the question which the experiment was intended to meet. This answer should state which swing, the long or the short, if either, takes the longer time, and whether the difference in time is large or small compared with the time of either swing.

We will now consider—

(2) How does the time required for a single swing depend upon the length of the *pendulum*, from the support down to the centre of the ball?

The teacher, holding a watch in his hand, draws ball No. 2 several inches aside from its position of rest and, releasing it at a convenient moment, gives a signal to the class, and the pupils count the number of single swings till, at the end of 20 seconds from the start, a signal is given to stop counting.

In a similar manner the number of swings of No. 3 in 20 seconds, and the number of swings of No. 4 in an equal time, are found, and the observations for the three pendulums are recorded in a table, as follows:

Pendulum.	Whole time.	Number of swings.	Time of one swing.	Length of pend.	Square root of length.
No. 2	20 sec.	....	....	36	6
" 3	"	....	....	9	3
" 4	"	....	....	4	2

The numbers to fill the fourth column must be found from those in the second and third columns. A comparison of the fourth column with the sixth column will probably show that there is a close relation between the time of swing and the length of a pendulum.

Finally, a comparison of No. 1 and No. 5, set in motion at the same time and with the same width of swing, will show whether the time of swing depends much upon the nature of the suspended body.

It will doubtless be noticed that the *width* of swing of the lighter body diminishes more rapidly than that of the heavier one. This gradual loss of motion is due to the resistance of the air. The resistance is about the same for both bodies, if they have the same size, shape, and velocity, but a light body is more quickly stopped by a given resistance than a heavier body. This is the reason why one cannot throw an acorn or a piece of cork so far as one can a stone of the same size.

It has been said above that pendulums are used to control clocks, but many clocks and all watches are controlled by means of vibrating *springs;* for these, like pendulums, are very regular in their swings and so are good timekeepers. The controlling springs (see the " balance " of a watch) must not be confused with the much larger *driving* springs, or *"main* springs," which are used in watches and in most clocks of the present day.

CHAPTER VII.

LIGHT.

WE say that a lamp *gives*, or *gives out*, light. This is true. Light is something that comes to our eyes from any object and enables us to *see* the object from which the light comes. The light which most objects send to our eyes has come to these objects directly or indirectly from the sun or from a lamp, as we may know from the fact that if we take these objects into a dark room, where no light falls upon them, they do not send any light to our eyes, and so we do not see them.

Of course we see many things every day upon which neither the sun nor any lamp is directly shining. We see them by what is called " daylight." This, however, is sunlight, although it may not have come straight from the sun to the objects that we see lighted up by it. It may have gone from the sun to a mass of clouds, from the clouds to the surface of fields or streets or walls of houses, and from such surfaces into corners where the sun itself is never seen.

Light as it comes from the sun, or from most lamps, is of many different kinds, all blended together so that the eye does not distinguish one kind from another; but when this mixture of lights falls upon certain objects, pieces of glass called *prisms*, for instance, the mixture is broken up and we see the different *colors*.

EXPERIMENT.

Hold a glass prism (No. XXII) in the direct sunlight in such a position that light after passing through the prism will fall upon a white surface not in the direct sunlight,

Most objects upon which sunlight falls make some change in it by destroying (or, rather, changing into something else) some parts of the mixture, so that the light which leaves these objects and comes to our eyes is a different mixture from that which the sun sends directly to us, and, as we say, has a different *color*.

Thus the light which comes through a solution of sulphate of copper is blue, because the sulphate of copper has stopped the other parts of the light which entered it from passing through. The sulphate of copper has added nothing to the light. It has merely taken away a certain part of it. One can get from a druggist little packages of dyes of various colors, which, when dissolved in water, give beautifully colored liquids. (See No. XXIII.)

When we speak of the color of an object we mean the color of the light which that object sends to our eyes. We distinguish one object from another by sight, mainly by difference in color, but partly by difference in brightness.

An object which allows any particular kind of light to pass through it without perceptible loss is said to be *transparent* to that kind of light. If it allows all kinds of light to pass through it without loss, so that sunlight suffers no change in traversing it, the object is said to be *transparent* and *colorless*. Water and good window-glass come near being transparent and colorless.

Most objects take their color from light which does not *appear* to have passed through them, but to have been *reflected* from their surfaces. It is known, however, that in many cases this light really has penetrated a little distance into the object and has then come back, so that we really get light that has passed through a thin layer of the substance. Light which is reflected from the *outer* surface of bodies is usually not changed in color by this reflection.

EXPERIMENT.

Let a beam of bright sunlight fall *very obliquely* upon a deep-blue solution of sulphate of copper, Fig. 52, or upon a plate of colored glass (see No. XXIV), and then pass by reflection to a white wall. Compare the color

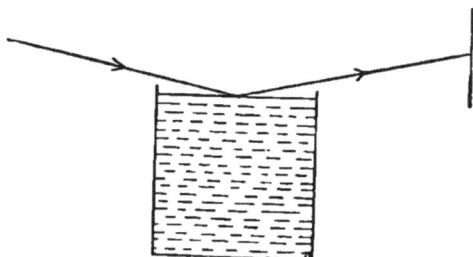

Fig. 52.

of this reflected light with that reflected at the same time and in the same way by an ordinary mirror (No. 22).

The reflecting surface which we make use of in a common mirror is not the front surface of the glass, but the metallic surface at the back. The glass is merely a convenient transparent support for the metallic layer, keeping it in shape and protecting it from being tarnished, as it soon would be if exposed to the air.

When we place an object in front of such a mirror and stand in a proper position we see an image, or "reflection," of the object, and we say that we see the object, or its image, *in* the mirror. If M, Fig. 53, is the mirror, O a point of the object, and P_1, P_2, P_3, and P_4 are the positions of four eyes, all may see at the same time an image of the point O in the mirror. Our first exercise in light is intended to answer the question whether all these eyes see the *same* image, that is, whether all are looking toward the same

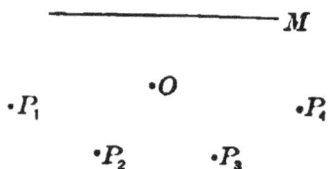

•P_1
•O
•P_4
•P_2 •P_3

M

Fig. 53.

point, and if so, where this point is—in front of the mirror, or behind it, or at its surface.

EXERCISE 19.

IMAGES IN A PLANE MIRROR.

Apparatus: A mirror (No. 22). A rectangular block (No. 9). Two rubber bands to hold the mirror to the block. Two straight-edged wooden rulers (Nos. 23A and 23B). A measuring-stick (No. 3). A sheet of thin white paper about 12 inches by 20 inches. A small block (No. 24). Attach the mirror to the large block by means of rubber bands in the manner shown by Fig. 54,

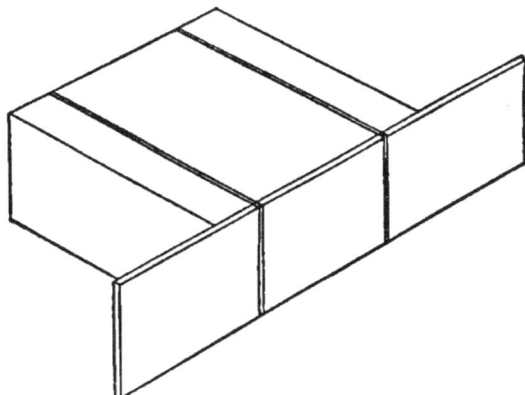

Fig. 54.

taking care that there shall be no twist in that part of each band which lies beneath the block, for such a twist causes the block to rest unsteadily and be easily moved out of place.

Draw a straight pencil-mark across the sheet of paper at its middle, and set the *back* surface of the mirror directly over and parallel to this line, the middle of the mirror being very near the centre of the sheet. See Fig. 55.

Draw on the sheet of paper in front of the mirror a triangle, each side of which should be several inches long, and no corner of which should be less than three inches from the mirror. It is well to have one angle of the triangle not directly in front of the mirror, but somewhat to one side, like point No. 1 in the figure.

Place the small block in such a position that the vertical

pencil-mark which it bears shall be directly over point No. 1 of
the triangle. Now lay a straight-edged ruler
upon the paper in such a position that one of
its long horizontal edges shall point directly
toward the image of the vertical pencil-mark,
as seen in the mirror. (Many persons cannot
do this at first unless they are especially instructed. Let the line along which the pupil
is to *sight* be *PQ* (Fig. 56), *P* being the point
nearer the eye. A person who is not near-
sighted should hold his eye eight or ten inches
distant from *P*, and should then direct the ruler
in such a way that the *point P*, the *point Q*, and
the image of the vertical pencil-mark seen in
the mirror, may all lie in one straight line. Do
not try to look along the vertical *side* of the
ruler, but hold the eye high enough to see all
the time the top of the ruler.) Then with a well-

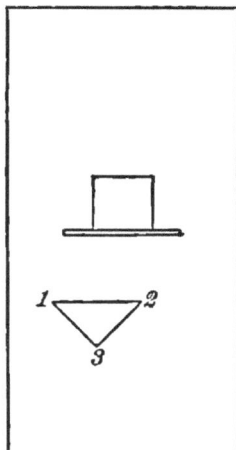

Fig. 55.

sharpened pencil draw upon the paper a fine clear mark alongside
that edge of the ruler which lies just beneath the line *PQ* (Fig. 56)
along which the *sight* has been taken. Mark this line 1, because it

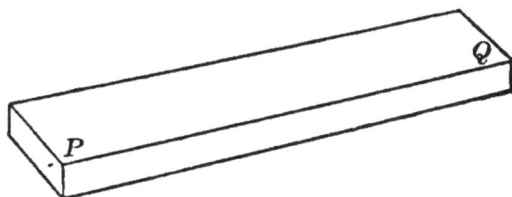

Fig. 56.

points toward the image of the vertical pencil-mark when this
mark is over point No. 1. Next, without disturbing anything else,
place the ruler in a new position, well removed from the position
just occupied, sight as before, draw another line alongside the
ruler, and mark this line also 1. Then with the ruler in a new
position, about half-way between the first two, if this is convenient, draw a third line in the same way and mark this also 1.
All this time the small block has remained unmoved and the
pencil-mark upon it has pointed straight down at point No. 1.

Now place the small block so that the pencil-mark will point

straight down at point No. 2. While it is in this position draw
three straight lines toward the image and mark each one of these
2.

Finally, put the pencil-mark over point No. 3, draw three
straight lines toward its image, and mark each of them 3.

(While drawing all these lines the pupil should look frequently
to see whether the back of the mirror remains in place. It may
be thrown out of place by a little blow or by rubbing the paper
hard to remove pencil-marks.)

When the three sets of lines have been drawn, the two blocks
and the mirror are removed from the paper, and each line is then
lengthened until it crosses both the others of the same set; that is,
each No. 1 line is continued toward or beyond the mirror till it
crosses the two other No. 1 lines. Then the No. 2 set and the No.
3 set are treated in the same way. (If a line has to be extended
far it is well to use two rulers, *A* and *B*, as shown in Fig. 57. First

FIG. 57.

A is put into position and a line is drawn alongside it. Then,
while *A* remains unmoved, *B* is carefully brought close to it, as
the figure shows ; then *B* is held firmly in place while *A* is pushed
forward to the position indicated by the dotted lines. *B* is then
removed without disturbing *A*, and again a line is drawn alongside
A. In this way a line may be continued nearly straight for a considerable distance.)

After each set of lines has been extended in this way, it will be
in order to answer the question whether all the lines of any one
set lead to the same point or nearly so, and, if so, where is this
point situated with respect to the mirror and to the point whose
image it is.

If the image of each point, No. 1, No. 2, and No. 3, can be thus
found, connect the image-*points* with each other by straight lines
and thus make an image-*triangle*.

Then fold the sheet of paper carefully along the pencil-mark by
which the mirror was placed, and holding the folded sheet against

a window, so that the light from without will shine through it, compare the size and shape of the two triangles and their relative positions with respect to the line along which the paper is folded.

The general rule for placing the image of any point should be recorded when it is found.

The final result aimed at in this exercise should be to enable the pupil to tell, without further experiment, in any new case given him, Fig. 58, for instance, in which $A B$ is the line upon which a mirror stands, the position of the image of points No. 1, No. 2, No. 3, and No. 4, and so the shape and position of the image of the figure at the corners of which these points lie.

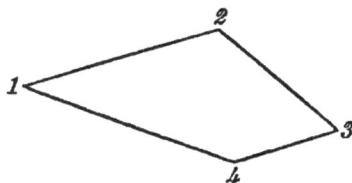

Fɪɢ. 58.

Suggestions for the Lecture-room.

On the basis of the pupils' work in the laboratory, which has shown where to locate the image I of a point O placed

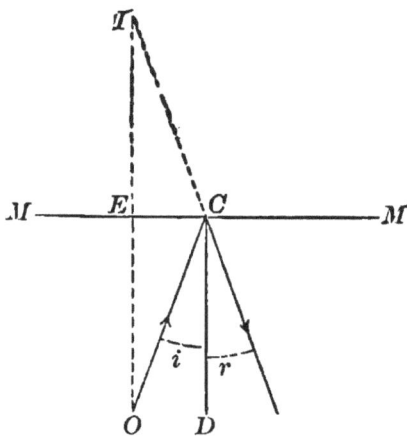

Fɪɢ. 59.

in front of a mirror MM, Fig. 59, prove that the "*angle of incidence*," i, which a ray from O striking the mirror at C makes with the line CD drawn at right angles with the mirror surface, is equal to the "*angle of reflection*," r, made by the same ray with the same line CD after reflection.

The line of proof is as follows: *Angles at E are right angles; $EI = EO$; EC is common to two triangles;*

hence triangle CEI is similar to triangle CEO. Then angle i = angle EOC = angle EIC = angle r.

We have not yet defined an *image* in strict terms. We now see that *the image I* (Fig. 60) *of a point O, placed be-*

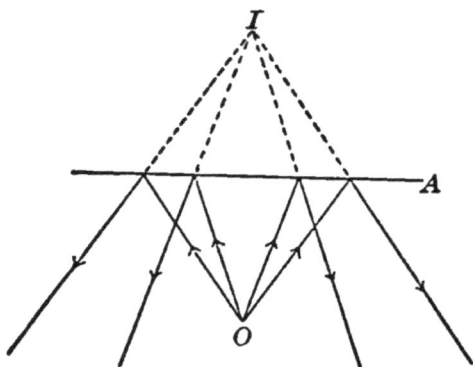

Fɪɢ. 60.

fore a plane mirror, A, is the point from which the rays of light that go from O to the mirror appear to diverge after reflection.

As the rays really do not come from I after reflection, but only *appear* to do so, this kind of image is said to be *unreal*, or *apparent*. It is often called a *virtual* image. We shall by and by find images that are *real*, images that will show on a white paper or cloth placed in the right position.

If any of the rays from O (Fig. 61) after reflection from the mirror A fall upon a second plane mirror B, they will be treated by this second mirror just as if they really came from I_1; that is, we shall, looking into the mirror B in the right direction, see an *image* of the *image* I_1, and this second image, I_2, will appear just as if it were the image of an actual object, sending rays from I_1.

The rays reflected first from A and next from B might then fall upon a third mirror, and give an image of the image I_2, and so on; but at each reflection there is some loss of light, and an image formed after many reflections might be dim.

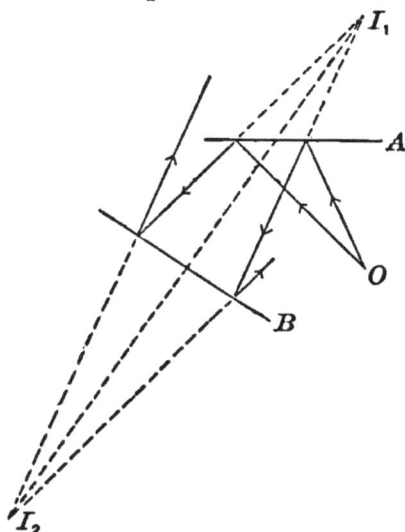

Let us consider the images of a point O formed

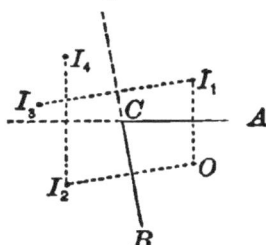

FIG. 61.

FIG. 62.

by two plane mirrors A and B (Fig. 62), making with each other an angle slightly less than 90°. We shall have one image, I_1, formed by mirror A without any help from mirror B. We shall have another image, I_2, formed by mirror B without any help from mirror A. There is also I_3, the image of I_1 seen in B, and I_4, the image of I_2 seen in A. We cannot with this arrangement of the mirrors get images of I_3 and I_4, for rays leaving mirror A as if diverging from I_4 would not strike either mirror again, and rays leaving mirror B as if diverging from I_3 would not strike either mirror again.

If the angle between the mirrors is made sufficiently acute, however, the number of images to be seen may be greatly increased. All the images appear to be as far from the corner C as the original point O is; that is, the point O and all its images are ranged on the circumference of a circle whose centre is at C.

EXERCISE 20.

COMBINATION OF TWO PLANE MIRRORS : KALEIDOSCOPE.

Apparatus: Two mirrors (No. 22). Two blocks (No. 9). A straight-edged ruler (No. 23). A paper protractor (No. 25). Four rubber bands. A sheet of white paper.

To study further the effect of combining two mirrors, and to see

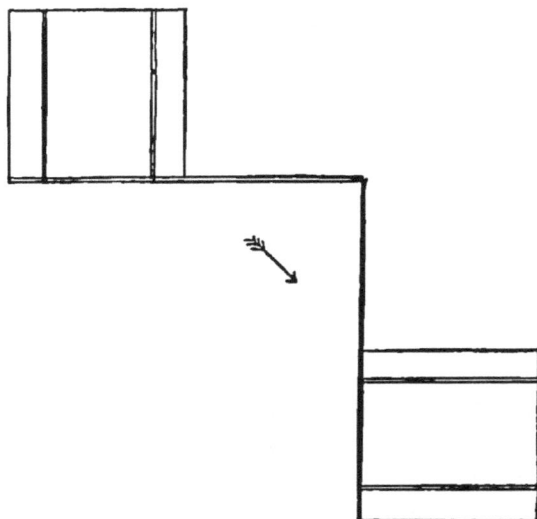

Fɪɢ. 63.

how the number of images formed in any given case depends upon the angle between the two mirrors, proceed as follows : Lay

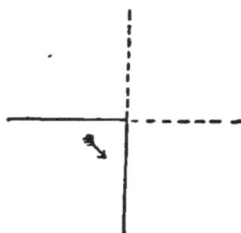

Fɪɢ. 64.

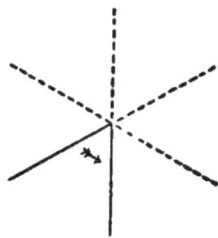

Fɪɢ. 65.

off on the sheet of paper an angle of 90°, and about one inch from the apex of the angle draw upon the paper a short, wide arrow

(Fig. 63). Then place the two mirrors so that their reflecting surfaces will be just over and parallel to the two lines inclosing the angle, supporting the mirrors as in Fig. 63.

Looking into the mirrors, count the whole number of images of the arrow visible in both of them, and then draw a diagram representing the arrow and its images in their proper positions with respect to each other, beginning in this way—Fig. 64.

Do the same thing with an angle of 60°, if time permits, beginning the final diagram as in Fig. 65.

(Two pupils will have to work together in this exercise unless two mirrors can be supplied to each pupil.)

Suggestions for the Lecture-room.

Velocity of light as found by observations on Jupiter's satellites and by Fizeau's method. (Consult almost any college text-book of general physics.)

Exhibit the camera obscura (No. XXV) described in the Appendix, and show how it is made.

Exhibit apparatus for the next Exercise, and define *convex, cylindrical,* and *centre of curvature* (centre of circle, see below).

EXERCISE 21.

IMAGES FORMED BY A CONVEX CYLINDRICAL MIRROR.

Apparatus : The mirror and its support (No. 26). A measuring-stick (No. 3). Small block (No. 24). Rulers (No. 23 A and B). Sheet of white paper. Also the plane mirror (No. 22) and its supporting block (No. 9).

Hold the board carrying the mirror horizontal, and look at the image of your own face in the convex surface of the mirror. You will see that the image is distorted, appearing too narrow for its length. Hold the board vertical and the image will be distorted in the opposite way, appearing too wide for its length. The object of the following experiments is to give you a better understanding of these curious effects.

Set the mirror-base on the table and bring one end of the plane mirror close to the surface of the curved mirror, as in Fig. 66. Then place the small block in front of both mirrors, as in Fig. 56, in such a position that you can see the block reflected in both mirrors at the same time.

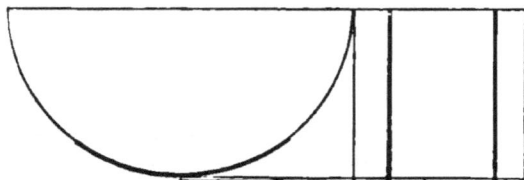

Do the two images thus seen appear of the same height?

Do they appear of the same width? Fill out, if you can, the following statement: *Lines of the object which are parallel to the straight lines of the cylindrical mirror appear ......*
.... in the cylindrical mirror
.... plane mirror.

FIG. 66.

Lines of the object which are at right angles with the straight lines of the cylindrical mirror appear .... in the cylindrical mirror .... in the plane mirror.

Place the mirror-base on the sheet of paper and draw around it a pencil-mark, marking the point *C,* Fig. 67, as the *centre of curvature.* About 5 cm. from the front of the mirror draw an arrow 6 cm. long, marking the ends and the middle as in Fig. 67. Then place the small block so that the vertical pencil-mark which it carries will point straight down at point No. 1.

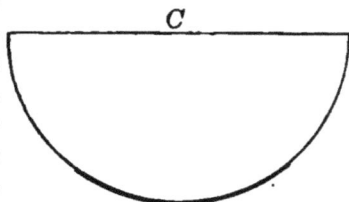

With the straight-edged ruler draw two lines, well apart, toward the image of this vertical line as seen in the mirror, avoiding parts of the mirror, if there are such, that do not give a good image of the line. Mark each of these lines 1. Then draw two lines for point No. 2 and two for point No. 3, in the same way.

FIG. 67.

Then clear the paper and prolong each pair of lines till it comes to a crossing-point. The three points thus found will locate the images of object-points No. 1, No. 2, and No. 3, respectively, and a line connecting these three image-points will give an idea of the shape of the image-arrow, whether it is straight or not, and whether its curvature, if it has any, is in the same general direction as the curvature of the mirror or in the opposite direction.

Draw a straight line from each marked object-point to the corresponding image-point, and prolong these three lines until they cross each other. Note where the crossing occurs.

Is the image longer or shorter than the object? Is it nearer to, or farther from, the mirror than the object is?

(It must be understood that the pupil is asked these questions only in regard to the particular case that he has tried. He cannot tell without further experiments or further instruction whether the answers he gives in this case would be true for all cases of objects reflected in mirrors such as he is using, for he does not know that the distance of the object from the mirror may not decide all these questions. The fact is, however, that if he has found correct answers to the questions asked for his one case, the same answers will be true for the same questions in all cases with *convex* cylindrical mirrors. The effects seen with *concave* mirrors are much more complicated.)

Suggestions for the Lecture-room.

With curved mirrors, as with plane mirrors, the law *angle of incidence = angle of reflection* holds. With the help of this law we can see why the image of a point is nearer the mirror, when this is convex, than the point itself is.

Let *O*, Fig. 68, be the object-point in front of the convex mirror *MM*, the centre of curvature being at *C*. A line drawn from *C* to any point of the mirror is at right angles with the mirror at the point of crossing.

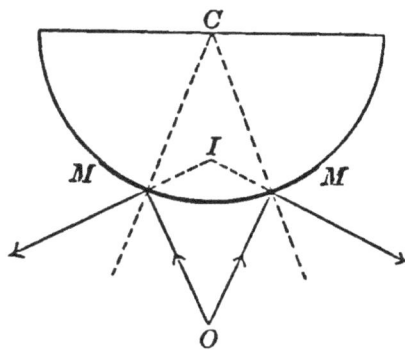

FIG. 68.

right angles with the mirror at the point of crossing.

Two rays going from *O* to the mirror-front appear after reflection to come from *I*, which is nearer the mirror than *O* is.

If the *concave* side of the mirror were used, it is easy to see from Fig. 69 that rays from a point *O* near the mirror-front would after reflection appear to come from a point *I*, which is *farther* from the mirror than *O* is.

It is evident that the rays

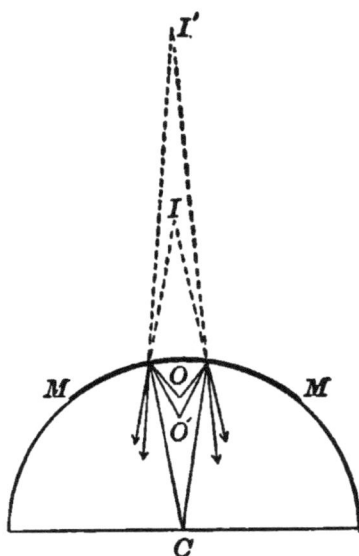

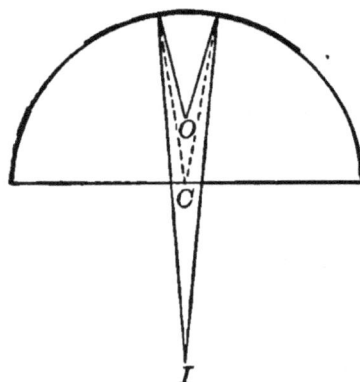

FIG. 69.   FIG. 70.

from *O* are more nearly parallel to each other after reflection than before.   Rays from a point *O'*, somewhat farther from the mirror than *O*, appear after reflection to come from a still more distant point, *I'*, and these rays are nearly parallel after reflection.   It is easy to see that if the object-point were put somewhat farther still from the mirror, the rays proceeding from it might, after reflection, be parallel to each other.   They would appear to come from a point as far as possible behind the mirror.

If the object-point is placed still farther away from the mirror, as at *O* in Fig. 70, the rays will after reflection be actually *converging*, and will cross at a point *I* in *front* of the mirror. This image I is a *real* image (see remarks following Exercise 19), and if O is bright enough, the image I may be seen, like a picture, on a piece of white paper or cloth placed in the right position.

If the object-point were placed where I now is, the image-point would fall where O now is. We see that the centre of curvature C lies between the object-point O and the image-point I in these two cases. This is always so in the case of *real* images formed by concave mirrors, unless the object-point is at C, in which case the image-point also falls at C.

EXERCISE 22.

CONCAVE CYLINDRICAL MIRROR.

Apparatus: The same as in the preceding Exercise, and in addition a common pin and two wooden toothpicks (or any two straight, slender objects 3 or 4 inches long).

From two points about 4 cm. apart on the mirror-front draw two radii, r r, on the base-board supporting the mirror, as in Fig. 71, extending them to the centre of curvature C. Draw arrow A about 0.8 cm. distant from the mirror-front and B about 1.5 cm. distant. Draw D about 1.5 cm. distant from the centre of curvature C. (All of this work should be done for the pupils before the class-work begins.)

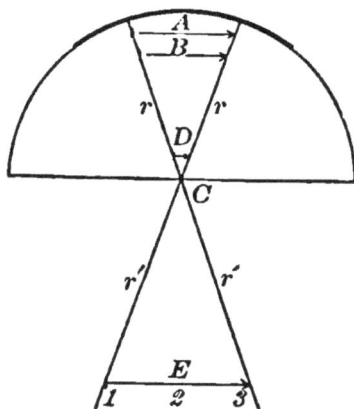

FIG. 71.

Let the pupil, holding the mirror about 10 inches from his eye, look at the images of A, B, and D.

Do the images of A and B point in the same general direction, from left to right in the figure, as the arrows themselves?

Is the same answer true of D and its image?

Are the images of A and B longer or shorter than the arrows themselves?

At the centre of A stand the pin upright, and, laying the two toothpicks on the base-board, point them toward the image of the pin, contriving to have them make a considerable angle with each other. In this way the position of the image is located. Is it behind the mirror or in front?

Is it, then, a *real* image or an unreal one?

By the same method locate the image of the pin when erected at the centre of B and when at the centre of D, asking and answering in each case the same questions that were asked when the pin was at the centre of A.

If time permits, continue the Exercise as follows:

Place the mirror on a sheet of paper and extend the two radii, r r by the lines r' r' drawn on the paper, as in Fig. 71. Draw the arrow E, 6 or 8 cm. distant from C, marking points 1, 2, and 3 upon it. Locate the image of each of these points by the method used in the preceding Exercise with the convex side of the mirror, drawing upon the paper the lines of sight and the image of the arrow.

Suggestions for the Lecture-room.

Curved mirrors made for common use are usually parts of *spherical* surfaces. The effects produced by such mirrors are really simpler than those produced by cylindrical mirrors.

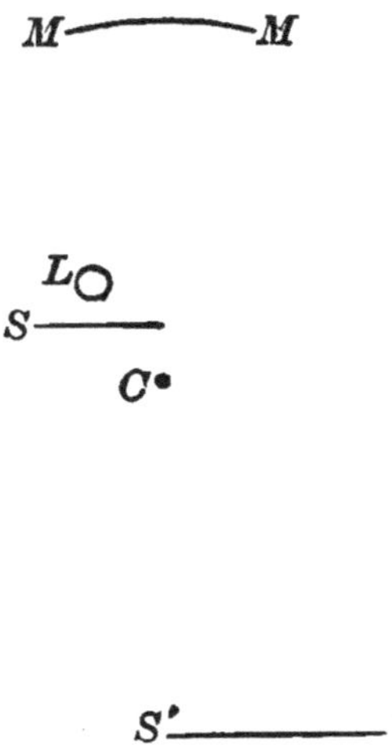

With a concave spherical mirror 5 or 6 inches wide (No. XXVI) interesting lecture-table experiments may be made in a slightly darkened room, the image of a candle-flame or, better, gas-flame being thrown upon a screen so as to be visible to all in the room. The screen should be of tracing-cloth or oiled paper, so that the image upon it may be seen from both sides. An opaque screen should hide the flame itself from the eyes of the pupils. Fig. 72 suggests a good arrangement, MM being the mirror, C its centre of curvature, L the

FIG. 72

flame, *S* the opaque screen, and *S'* the tracing-cloth screen.

The positions of *L* and *S* may be greatly varied and may be interchanged, but the least distance of either from the mirror should be rather more than one half the radius of curvature of the mirror, if *real* images are desired.

CHAPTER VIII.

REFRACTION.

In the experiment made with a prism the class may have noticed that the light did not go in the same direction after leaving the prism as before entering it. Some members of the class in looking into pools or vessels of water may have noticed that objects beneath the surface are not exactly where they seem to be.

EXPERIMENT.

Place a straight stick in an oblique position, partly in and partly out of water. Exhibit this to the class in several aspects, showing the apparent bending or disconnection of the stick at the surface of the water.

These curious effects are due to the fact that when light goes obliquely from air into water or glass, or any other transparent body, or when it comes out obliquely from any such body into air, it suffers a change of direction just at the surface of the body. This change of direction is called *refraction.*

The amount of the bending, or refraction, which a ray of light suffers at any surface depends partly upon the two substances which meet at this surface, and partly upon the angle, i, Fig. 73, which the ray makes with a line NN, which is at right angles with the surface at the point C where the ray strikes the surface.

If the space above the line AB represents the air-space, and that below this line the water, or glass, or whatever substance it may be that lies there, solid or liquid, the course of the ray is changed at the surface in such a way that the angle r which it makes with NN inside the solid or liquid is smaller than the angle i.

The angle i in Fig. 73 is called the *angle of incidence*. The angle r is called the *angle of refraction*.

If the ray were represented as coming in the opposite direction, that is, first along R and then along I, r would be the angle of incidence and i would be the angle of refraction. The ray would be bent just as much at the

Fig. 73.

surface as it is when going first along I and then along R.

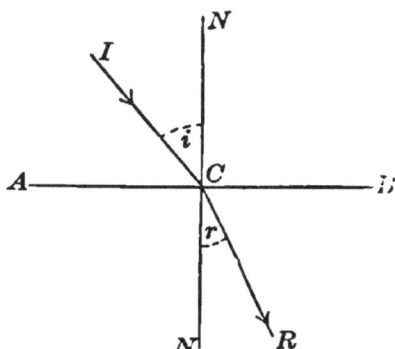

When the direction of I is changed the direction of R is changed. The way in which the change of one depends upon the change of the other is easily shown by means of Fig. 74. I, I', and I'' show three rays all of which come to the point C and then separate, the first going along R, the second along R', the third along R''. The circle whose centre is at C is drawn with any convenient length of radius. The dotted lines, n, n', n'', and m, m', m'', are drawn from the points where the rays cut the circumference to the line NN at right angles.

If this figure has been drawn so as to accord with the results of experiments on light rays, we shall have

$$\frac{n}{m} = \frac{n'}{m'} = \frac{n''}{m''},$$

and any one of these equal ratios is called the *index of refraction from* the substance or *medium* through which the ray comes, *to* the substance or medium into which it goes,

If now we can measure $\dfrac{n}{m}$ in any given case, we shall have a quantity which is very useful in physics, for by means of it we can *calculate* at once the value of a new m

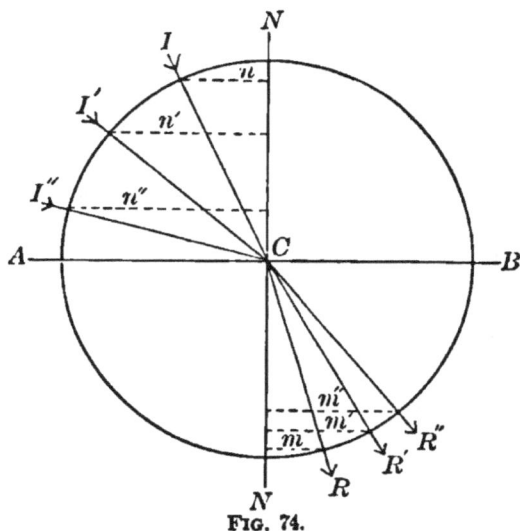

FIG. 74.

to go with any new n; that is, we can, if we know the angle which any ray makes with NN in one medium, find without further experiment the angle which the same ray makes with NN in the second medium. Exercise 23 shows how to find the ratio $\dfrac{n}{m}$ for the case of air and water.

EXERCISE 23.

INDEX OF REFRACTION FROM AIR TO WATER.

Apparatus: Articles Nos. 3, 14, 15, 23A, 23B, 27, 28, and a sheet of paper about 6 inches square.

Put the partition N in place, as shown in Fig. 75, and pour water into the jar until its surface comes within 1 or 2 mm. of the middle tooth of the partition. Then by means of the plunger (No. 14), attached to the side of the jar by means of its clasp, raise the level of the water till the apparent distance between the middle

tooth of the partition and its reflection in the water surface is less than 1 mm. (To see this reflection well, one should look through the wall of the empty part of the jar.

Then the brass index b is attached to the jar, as shown in Fig. 75, and is raised or lowered until an eye on the line Cg, 8 or 10 inches from the jar, can barely see p, the very tip of b, apparently in a straight line with Cg. This setting should be made with care, and *after* it is made the experimenter must look to see whether the tooth at C is clear of the water. If its lowest edge *touches* the water the setting is useless, and all of the adjustments must be made anew before a *reading* is made.

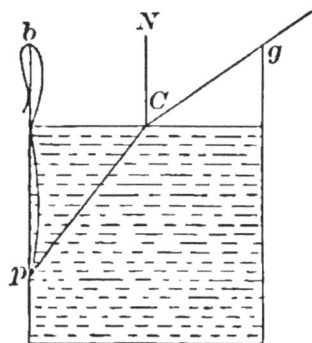

Fig. 75.

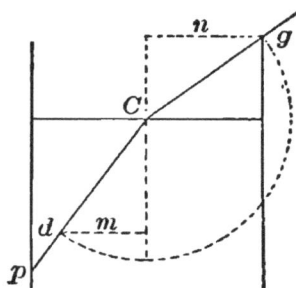

Fig 76.

When all the adjustments have been successfully made, measure carefully the distance from the top of the jar down to the tip p of the index, the measuring-stick being kept outside the jar. (The tip p should be close against the glass).

Measure now the inside diameter of the jar.

Measure also, unless it is already known, the distance of C below the top of the jar. (It is well to have this distance, which is somewhat troublesome to measure accurately, given by the teacher. Partitions of different depth might be used in order to vary the angles of incidence and refraction.)

Now make a drawing, of full natural size, of the sides of the jar, the water surface and the partition, as in Fig. 76, continuing the partition line, by means of dots, well down into the jar. Put p in its proper place, and then draw the lines pC and Cg.

Now with C as a centre and with any convenient radius, Cg for instance, one may draw a circle cutting CP at d. Then, since pC is the course of a ray of light in the water, and Cg the course of the same ray after leaving the water, the *index of refraction* from *water* to *air* is $\dfrac{m}{n}$, while the index of refraction from *air* to *water* is $\dfrac{n}{m}$. (It is customary to state the index of refraction *from* air *to* water.) It is evident that the *circle* need not be drawn. It is just as well to find the point d by measuring off Cd equal to Cg.

If the jar used in this Exercise is not pretty level at the top, or if the partition is not just at the middle of the jar, it is well, after making one setting of the index and one measurement of its position, to turn the jar about, transferring the index to the other side, and make a new setting and a new measurement. The *mean* of the two measurements thus made should be nearly free from any error caused by irregularity of the jar or of the partition's position.)

Suggestions for the Lecture-room.

If light goes through a transparent plate whose sides are parallel to each other, the refraction at the second surface *undoes* that which happens at the first surface, and the

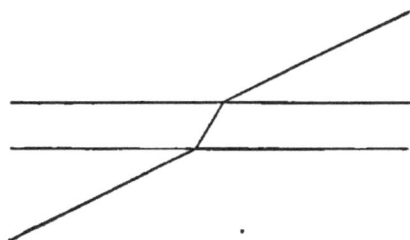

FIG. 77.

light has the same direction after leaving the plate as before entering it. See Fig. 77. Light passing through a *prism* (see Fig. 78), usually suffers at the second surface a further bending away from its original direction. The separation of colors by which the

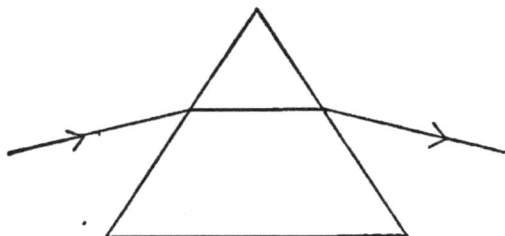

FIG. 78.

prism produces what is called a *spectrum* is due to the fact that the different kinds of light are bent different amounts by the prism. Blue or violet is refracted most and red least.

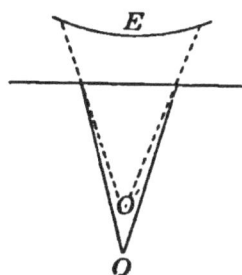

FIG. 79.

When we look straight down into water at any small object it appears to be in its true direction from the eye, but nearer than it really is. Fig. 79 indicates why this is so. *O* represents the object, *E* the eye, and *O'* the apparent position of the object.

LENSES.

A *lens* is, usually, a piece of glass whose two faces are parts of spherical surfaces.

Sometimes there is a cylindrical surface between the two spherical faces.

Fig. 80 shows various lenses as they would look if cut through the middle.

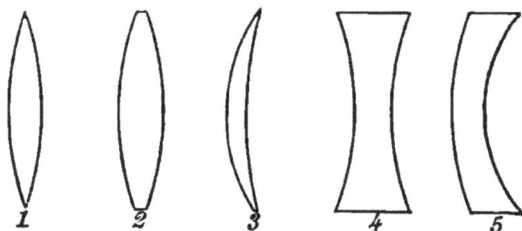

FIG. 80.

The lenses we shall use will be much like No. 1 in this figure. The two sides are supposed to be just alike.

To understand such a lens better we will make use of Fig. 81.

C_1 is the centre of the spherical surface of which *ASB*

is a part.   It is called the *centre of curvature* of the face
ASB.   C₁ is the centre of curvature of the face *ARB.*

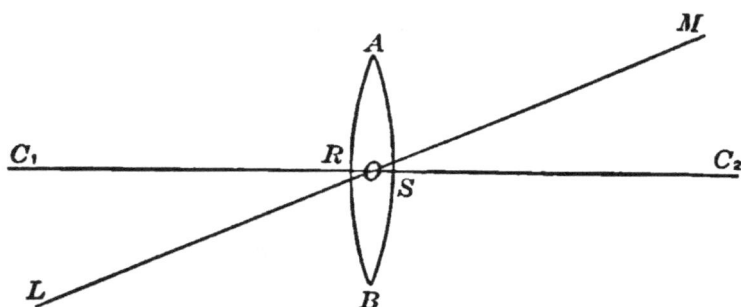

Fig. 81.

The straight line C_1OC_2, continued to any distance in
each direction, is called the *principal axis* of the lens.

Any straight line going, like *LM,* obliquely through the
centre of the lens is called a *secondary axis* of the lens.

If the two faces of a lens are exactly alike, as we suppose
them to be here, any ray of light going through the centre
of the lens, the point *O,* will have the same direction after
leaving the lens as before entering it, because the two little
spots of surface at which it enters and leaves the lens are
parallel to each other, so that the ray is affected just as if
it were going through a plate with parallel faces.

Rays entering a lens, Fig. 82, *parallel to its principal*

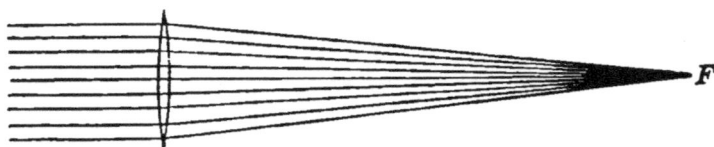

Fig. 82.

axis are refracted in such a way that after leaving the lens
they will cross this axis. They do not all cross at one
point, but if the faces are near together, and are *very small*

parts of spherical surfaces, as in our lenses, such rays will cross at or near a certain point, *F*, on the principal axis, and this point is called the *principal focus* of the lens.

The distance from the principal focus to the nearer face * of the lens is called the *focal length* of the lens.

Focal length is a quantity of very great importance in dealing with lenses, and the next Exercise will show how to find it by experiment. For this purpose we need to have the light come to the lens in rays nearly parallel to each other and to the principal axis. This we can do by taking the light from any small spot of any distant but distinct object; for instance, a chimney or a church spire outlined against the sky.

EXERCISE 24.

FOCAL LENGTH OF A LENS.

Apparatus: The lens mounted on a block (No. 29). A meter-rod (No. 2). A small block (No. 20) bearing a white cardboard screen (No. 30). A common pin.

FIRST METHOD.

Place the lens and the screen upon the rod, as in (Fig. 83), and

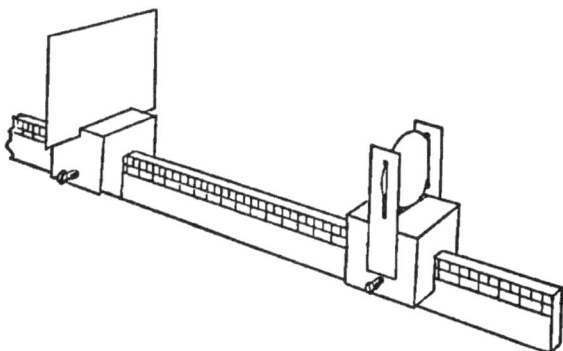

FIG. 83.

point the rod at some distant object seen against the sky in such a way that the light from this object will pass through the lens

---

* See Appendix VIII. of Hall and Bergen's Text-book of Physics.

and then fall upon the screen. Move the screen back and forth until that part of the image which lies on or near the principal axis of the lens is made as distinct as possible. Then by means of the graduations on the meter-rod, or by an independent measuring-stick if this is preferred, note the distance from this part of the image to the nearer face of the lens. This is the *focal length*. (The image is formed because light coming from any one small spot of the object is brought to a small spot again by the lens. The image is made up of such small spots each in its own place. For the purposes of this class the *distant* object need not be more than 30 or 40 feet from the experimenter. The images on the screen will be much more distinct if the apparatus is used in the back part of the room, well away from the windows.)

SECOND METHOD.

Remove the screen from its block and put the pin upright in its place. Let the pin, thus mounted, be placed on the meter-rod, about as far from the end of the rod as the pupil usually holds a book from his eyes when reading. Place the lens somewhat farther from the same end of the rod.

Place the eye at this end of the rod and, looking sharply at the *pin*, direct the rod and adjust the lens in such a way that the light from some distinct distant object will pass through the lens and form an image *in the air* close to the pin. To decide whether the image is nearer the eye than the pin is, move the eye to and fro, to the right and the left, watching the pin and the image. If the pin is more distant than the image, it will, when the eye is moved toward the right, appear to move across the image toward the right. If the pin is nearer than the image, it will, when the eye is moved toward the right, appear to move across the image toward the left. The rod should not be held in the hands during this test, but should be placed on some steady support. (To see the reason of this test, close one eye and hold the two forefingers, some inches apart, in line with the other eye, so that one finger hides the other. Then move the eye to the right and left, and notice the apparent movement of the fingers with respect to each other.) Continue the adjustments until the test described fails to show which of the two, the pin or the image, is nearer the eye. Then measure the distance from the pin to the lens. It should be the focal length of the lens.

Compare the values of the focal length given by the two methods.

The second method is more difficult, but it gives, perhaps, more accurate results, and it can be used in cases where the image is too faint to show clearly upon the screen.

(Each lens should be numbered on a bit of paper pasted upon it. Each pupil should record the number of the lens he uses. The teacher should know the focal length of each lens.)

Suggestions for the Lecture-room.

Were the images seen in Exercise 24 real or virtual ?

Were they *erect* or *inverted;* that is, were they right side up or wrong side up ?

As there is an image *in the air,* the pupil may not see why this image cannot be seen by a whole class at once without the use of a screen. It is because the light forming the image in the air goes straight on *through* this image, and can be received only by placing one's self behind the image; while the light which forms an image upon a screen is by the threads of the screen reflected back in all directions, and therefore some part of it reaches every eye.

If a bright point were placed at the principal focus of a lens, what direction would the rays going from this point to the lens have after passing through the lens ?

In the next Exercise we shall ask at what distance from the lens the image is when the object is not a distant one.

EXERCISE 25.

RELATION OF IMAGE-DISTANCE TO OBJECT-DISTANCE:
CONJUGATE FOCI OF A LENS.

Apparatus: The same lens that was used in Exercise 24 (No. 29). A meter-rod. Block (No. 9). Small block (No. 20) with a cardboard screen (No. 30). Small kerosene lamp with a blackened chimney (No. 31).

(To economize space upon the laboratory-tables it will probably be necessary to have pupils work in pairs in this Exercise. Each pair should know or be told the focal length of its lens at the outset, so as to lose no time in beginning the Exercise.)

Arrange the apparatus according to Fig. 84. The cross on the chimney, lighted up by the flame behind, is the *object* whose

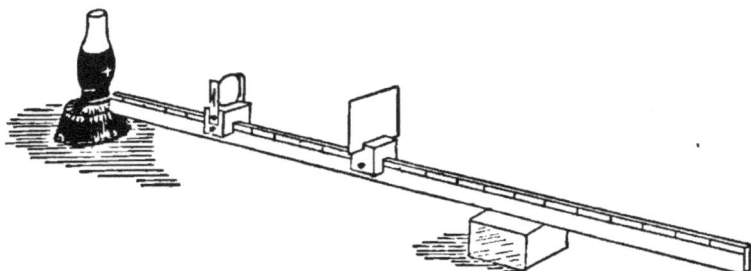

FIG. 84.

image is to be received upon the screen. One end of the meter-rod rests upon the base of the lamp, just beneath the cross.

Place the screen at first at a distance from the cross about equal to three times the focal length of the lens. Then move the lens back and forth on the rod between the cross and the screen, and see whether in any position it gives upon the screen a clear image of the cross. If it does, measure the distance from the lens in this position to the *cross*, and write this distance as the first number in a record-column headed D_o (object-distance). Measure also the distance from the lens to the screen, and put this distance as the first number in a record-column headed D_i (image-distance.)

If, with the present position of the screen and cross, there is *no* position of the lens that will cause a distinct image of the cross to fall upon the screen, move the screen one or two centimeters farther from the cross, and then try again to get a good image. If still none is found, move the screen still farther away, continuing the trial till a distinct image is obtained. Then measure and record the D_o and D_i as already described. (Very little time need be spent upon these first successive trials.)

Then at one move place the screen about 10 cm, farther still from the cross, find a position of the lens that will give a distinct image, measure and record D_o and D_i as before. Without moving the screen see whether there is any other position of the lens that will give a distinct image. If there is, measure and record the D_o and the D_i for this position of the lens.

Move the screen 10 cm. farther away, and then do exactly as before.

If there is time, move the screen two or three more times, adjusting the lens, measuring, and recording each time. It is better to make a moderate number of settings and readings well than a large number carelessly, but an error of one or two millimeters in these readings will be of little consequence.

Suggestions for the Lecture-room.

The distance from object to image in any case of Exercise 25 is $D_o + D_i$, and we may call this D_{oi}. This distance was shortest in the first case recorded. Let each member of the class divide the D_{oi} of this case by the focal length of his lens. Is the quotient found by any one as small as 3? Is it as large as 5?

When the screen was farther away, was there usually more than one position of the lens that would give a distinct image, the screen remaining unmoved?

If you were told that in a given case the D_o was 20 cm. and the D_i 60 cm., could you tell what the other possible D_o and D_i would be for the same positions of object and screen? Look at your record-columns for Exercise 25, and see whether they help you to answer this question.

Definition.—Two points so placed with respect to a lens that an object placed at either one of them will have an image at the other are called *conjugate foci* of the lens.

Let each member of the class call F the focal length of the lens which he used in Exercise 25, and let him test the truth of the formula

$$\frac{1}{F} = \frac{1}{D_o} + \frac{1}{D_i},$$

or, what means the same,

$$D_o \times D_i = F(D_o + D_i),$$

for all the cases which he tried and recorded in Exercise
25.

In the preceding Exercises the object presented to the
lens has been small, or has been at such a distance as to
give a rather small image. It is now desirable to study
larger images, and to study them with especial reference to
their *shape* rather than their distance from the lens. We
shall in the next Exercise find the shape and size of an
image of an arrow placed at right angles with the principal axis of the lens and not far from the lens. We shall
not attempt to find the whole image at once, but shall find
separately the images of several points of the arrow, and
then make an approximate image of the arrow by connecting these points.

EXERCISE 26.

SHAPE AND SIZE OF A REAL IMAGE FORMED BY A LENS.

**Apparatus: The lens (No. 29). Measuring-stick (No. 3). Block
(No. 20) carrying in the narrow slot on its top a piece of wire (No.**

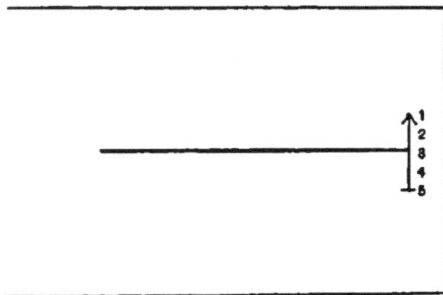

FIG. 85.

32) extending first horizontally and then downward
(see Fig. 85). A ruler (No.
23). Block (No. 24). A
sheet of paper about 1 ft.
wide and 3 ft. long, having
near one end an arrow 8 cm.
long, drawn at right angles
with a pencil-mark about
30 cm. long, and marked,
or numbered, as shown by
Fig. 85. Weights (No. 19)
to hold the corners of this

sheet in place on the table.

Arrange the apparatus as shown by Fig. 86, the centre of the
lens over a point on the long pencil-mark, at a distance from the
centre of the arrow about equal to 1½ times the focal length of the
lens, block No. 24 so placed that the vertical mark upon its face
points straight down to point No. 3 of the arrow. This vertical

mark will now cross the principal axis of the lens, if the lens is *accurately* placed.

Place the other block near the other end of the paper in such position that the vertical part of the wire it carries shall be near the principal axis of the lens. Keep the eye 8 or 10 inches distant from this part of the wire, on a level with the centre of the lens and in line with the centre of the lens and the vertical part of the wire. Look at this part of the wire so as to see it *distinctly,* and note whether you can see at the same time, near the wire the image of the pencil-mark on the farther block.

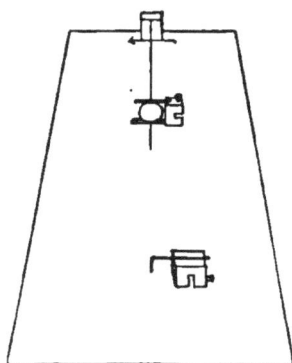

FIG. 86.

If so, find out by moving the eye to the right or left, as in **Exercise 24,** whether this image is more or less distant from the eye than the vertical wire is. Then move the block carrying the wire into such a position that the image and the wire seem to keep close together when the eye is moved a considerable distance to the right or left. When this adjustment is made, put a dot on the paper just beneath the vertical wire and mark this dot 3. It represents the image of object-point No. 3.

Find in a similar manner the image-points 1, 2, 4, 5, corresponding to the object-points 1, 2, 4, 5. The pupil must take care not to let any idea he may have as to the position where an image-point *ought* to be affect his judgment in deciding where it *is*.

After all the five image-points are found, connect them, No. 1 to No. 2, No. 2 to No. 3, etc., by means of straight lines, thus getting at least a rough representation of the whole image.

Draw from each object-point toward the corresponding image-point a straight line as long as the ruler (No. 23) and note the point where these lines cross each other.

Suggestions for the Lecture-room.

The formation of the image-points in Exercise 26 is illustrated by Fig. 87. One ray from the object-point A follows a secondary axis, passes through the centre of the lens O, and its direction after leaving the lens is the

same as before entering it. (Its direction *inside* the lens
is not quite the same, but the figure does not show this.)
Another ray from *A* runs parallel to the principal axis

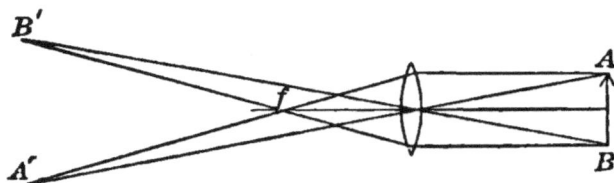

FIG. 87.

before entering the lens, and will therefore pass through
the principal focus, *f*, on the farther side of the lens. The
crossing of these two rays at *A'* shows the position of the
image of *A*.

In a similar way *B'*, the image of *B*, is located.

We see from Exercise 26 and from Fig. 87 that, when an
object-point is farther from a lens than its principal focus
is, the rays going from this object-point to the lens are
bent by the lens in such a way that, after leaving it, they
converge to a point again. We know that if the object-
point were placed at the principal focus the rays going
from it to the lens would emerge from the lens parallel to
each other.

It is, therefore, not difficult to see that, if the object-
point were placed *between* the lens and its principal focus,
the rays going from it to the lens would be divergent still,
after leaving the lens, though less divergent than before
entering it. In the next Exercise we shall have a case of
this kind.

EXERCISE 27.

VIRTUAL IMAGE FORMED BY A LENS.

Apparatus : The same as for the previous Exercise, except that
the sheet of paper need not be more than one half as long, and that

the arrow upon it should be 4 cm. long and about 20 cm. distant from one end.

Place the lens between the arrow and the nearer end of the sheet of paper, at a distance from the arrow about equal to ⅔ its focal length, and in such a position that its principal axis extends over the middle point of the arrow. Place the small block (No. 24) with vertical pencil-mark pointing straight down at the middle point, No. 3, of the arrow. Turn the vertical part of the wire on the other block so that it will point up instead of down, and place this block some inches behind the other one.

Holding the eye 8 or 10 inches from the lens, look *through* the lens at the image of the vertical pencil-mark and at the same time *over* the lens at the vertical part of the wire. Bring the wire into line with the image and then by the usual test find which of them is the more distant. Move the wire back and forth until it concides in position with the image. Then mark with a figure 3 the point just under the wire. This represents the image of object-point No. 3.

In a similar manner locate the images of points 1, 2, 4, and 5.

Connect the image-points by straight lines, from 1 to 2, from 2 to 3, etc., thus forming an image of the arrow.

Draw a straight line from each image-point to its corresponding object-point, and note where these lines will cross each other if continued.

Suggestions for the Lecture-room.

The images observed in Exercise 27 were *virtual* images. They could not be shown upon a screen, and were not formed by the actual crossing of light rays. Fig. 88 will serve to illustrate the way in which virtual image-points are formed.

Let AB be the object, placed between the lens LL' and the principal focus F'. To find the position of the virtual image of the point A, draw AI parallel to the principal axis of the lens. This ray will, after leaving the lens, pass through F, the principal focus on the farther side, and so will appear to have come along the path MF. Draw another ray, AC, passing through the *centre* of the lens. This ray will, after leaving the lens, have the same direction as before entering it and will be represented by the line CN. If, then, we carry back the line CN till it crosses the line MF', also carried backward, the point A', where the crossing occurs, is a point from which both of the rays appear to come. A' is, then, the virtual image of A. By a similar process B' is found to be the virtual image of B. P', the image of the point P, is here represented as lying in the straight

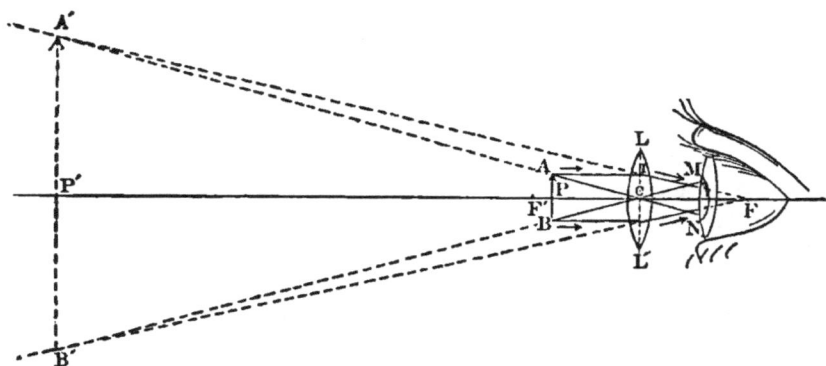

Fig. 88.

line between A' and B'. It is usually so represented in books. Exercise 27 shows that it does not lie there.

The image $A'B'$ is evidently larger then the object AB. Whenever a virtual image is formed by a convex lens, this image appears, to an eye placed in any ordinary position on the other side of the lens, larger than the real object would look if held at a comfortable seeing-distance from the eye. Hence the name *magnifying-glass,* so commonly given to a lens used as in Fig. 88.

We have seen that an *image* of an *image* may be obtained with mirrors. So the *image* formed by one lens may become the *object* for another lens.

In a common telescope the light from a distant object passes first through a lens, or combination of lenses, called the *objective*, and forms within the tube of the telescope a real image of the object. Then another lens, or combination of lenses, called the *eye-piece*, treats this image just as the lens LL' in Fig. 88 treats the object AB.

A microscope, like a telescope, consists of an objective and an eye-piece, the former giving a real image which the latter treats as an object. The objective of the microscope is of very short focal length, and the object to be examined is placed near the focus, so that the image which the objective forms is much larger than the object itself. The magnifying process thus begun is continued by the eye-piece.

APPENDIX A.

ALL the articles in the first list here given, except No. 31, should be furnished to each member of the laboratory section. Article No. 31 should be furnished to each pair of experimenters.

LIST OF ARTICLES REFERRED TO BY NUMBER IN THE "EXERCISES" OF THIS BOOK.

No. 1. A 10-cm. section of a meter-rod.

No. 2. A meter-rod marked on one side in feet and inches.

No. 3. A 30-cm. bevel-edged measuring-stick, marked on one side in inches.

No. 4. A wooden water-proofed cylinder about 8 cm. long and 4.5 cm. in diameter, loaded internally with shot so that it will float nearly submerged in water.

No. 5. A brass can about 14 cm. tall and 7 cm. in diameter, having a slightly declining, straight, overflow tube, about 6 cm. long and 0.8 cm. in internal diameter, extending from a point about 1.5 cm., clear, below the top of the can (see Fig. 8).

No. 6. A brass catch-bucket, with a wire handle, capable of holding about 175 gm. of water, and weighing not more than 50 gm.

No. 7. An 8-oz. spring-balance graduated to 0.5 oz. (The Franklin Educational Company of Boston offers an improved balance, graduated on one side in 10-gm. intervals and on the other side in 0.25 oz. intervals. It is,

FIG. 89.

moreover, especially adapted for use in the horizontal position. This improved balance is better for this course.)

No. 8. A rectangular water-proofed block of wood, about 7 cm. long and 4.5 cm. square on the end, so loaded internally with shot that it will sink in water, but not enough to make it weigh more than 225 gm.

No. 9. A rectangular water-proofed cherry block about 7.5 cm. × 7.5 cm. × 3.8 cm. This block should be smooth, and therefore the water-proofing should be done by soaking it in *very hot* paraffin. For the best results this soaking should be done in a vacuum. Excess of paraffin should be scraped off before the block is used.

No. 10. A one-gallon glass jar of *good* quality. (It is poor economy to buy a poor jar and have it break with a liquid in it.)

No. 11. A lump of roll sulphur weighing about 175 or 200 gm. It is not worth while to cast these lumps into regular cylindrical form.

No. 12. A lead sinker, with wire handle, weighing about 175 gm.

No. 13. A water-proofed wooden cylinder about 1 cm. in diameter and 20 cm. long. Doweling-rod, furnished by hardware dealers, serves well when water-proofed.

No. 14. A holder for keeping No. 13 upright in water. It consist of a water-proofed wooden rod about 12 cm. long and 1.3 cm. square on the end, provided with a clasp for attaching it to the side of a jar, and with two screw-eyes projecting from one side, the rings of which are large enough to let the cylinder No. 13 slip easily through them, but not large enough to allow the cylinder to tip far from the vertical position (see Fig. 18).

No. 15. A cylindrical glass jar, about 14 cm. tall and 10 cm. in diameter, with level top.

No. 16. A broad-mouthed bottle with ground-glass stopper, standing not much more than 11 cm. tall with stopper, and weighing, when filled with water, about 175 or 200 gm.

No. 17. A lever and supporting-bar. The lever is a 30-cm. section from a meter-rod, pivoted upon the *smoothed* cylindrical body of a brass screw which is driven horizontally into the end of a bar of hard wood about 25 cm. long, 5 cm. wide, and 3 cm. thick. A brass plate projecting from this bar and overhanging the middle of the lever prevents the lever from tipping far, while it allows sufficient freedom of motion. The lever itself, except for a distance of 2 cm. each side of the middle, is cut away so that its top is level with the upper part of the hole through the centre. There should be a hole about 0.5 cm. in diameter running downward through the middle of the supporting-bar. (Fig. 24.)

No. 18 (*A* and *B*). Two brass scale-pans about 6.5 cm. square, each with its suspending threads weighing accurately 1 oz. (that is, not differing from this weight by more than .01 oz.). Each pan is suspended by four strong linen threads meeting in a knot about 20 cm. above the pan, two of them continuing in a loop about 4 cm. long above this knot. (Fig. 24.)

No. 19. A set of iron weights, 8 oz., 4 oz., 2 oz., and two 1 oz., making a total of 16 oz. No weight should be in error more than .01 oz.

No. 20. A cubical block of wood about 3.7 cm. on each edge. A groove about 1 cm. wide and 2 cm. deep extends through the lower part of the block with the grain of the wood. An ordinary short screw extends through one side of the block into this groove, and serves to fix the block in position upon a meter-rod. *Across* the grain at the top of the block is a slot about 0.1 cm. wide and 0.5 cm. deep. (Fig. 83.)

No. 21. Two bits of wood each about 8 cm. long and 1 cm. square on the end, for supporting the spring-balance in a horizontal position. (Fig. 41.)

No. 22. A plate-glass mirror about 15 cm. long, 3.8 cm. wide, and 0.2 cm. thick, the coating on the back protected by paint or varnish.

No. 23 (*A* and *B*). Two straight-edged rulers of some wood that will keep its shape well,—white pine, for instance, —each about 30 cm. long, 5 cm. wide, and 1 cm. thick.

No. 24. A block like No. 20, but without the large slot and the screw. One side of this block is coated with white paper, and a vertical pencil-mark or ink-mark is made across the middle of this paper. (Fig. 86.)

No. 25. A Walter Smith "school square," or other equally good paper protractor.

No. 26. A cylindrical mirror of nickel-plated brass, about 4 cm. tall and 5 cm. wide, cut from seamless tubing 4 inches in diameter and $\frac{1}{16}$ inch thick, mounted upon a semicircular base-board of wood of the proper radius of curvature. The base-board should be about 1.5 cm. thick.

No. 27. A brass partition made to fit the small glass jar (No. 15), and to extend downward into the jar a distance equal to about one-third the diameter of the jar. It should be made of sheet brass about .07 cm. thick. The method

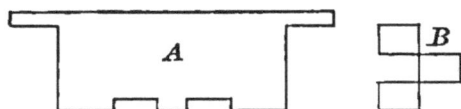

Fig. 90.

of shaping and adjusting the partition is suggested by Fig. 90, where *A* shows a side view, and *B* an end view, of the partition. The flanges shown in *B* are bent more or less in adjusting the partition to fit the jar closely, but without too much pressure.

No. 28. An index of thin sheet brass made to clasp the side of the jar (No. 15). This index is a strip about 15 cm. long, before bending, and 1 cm. wide, tapered to a point at one end. To enable it to clasp the jar, about 3 cm. at the untapered end is bent over. (See *pb* in Fig. 75.)

No. 29. A circular (not elliptical) double-convex spectacle-lens, having a focal length not less than 12 cm. and not more than 16 cm. This lens is mounted on a block similar to No. 20, and is held in place by two brass strips, each fastened to the block by a single screw and extending about 3.5 cm. above the top of the block. Each strip has a narrow vertical slot, cut by a circular saw, which does not extend to the top of the strip. The edges of the lens fit into these slots, and the lens is so held securely in an upright position (see Fig. 83). The strips may be about 1 cm. wide and .07 cm. thick.

No. 30. A white cardboard screen about 8 cm. square, of such thickness as to be held firmly in the narrow slot of the small block No. 20. (Fig. 83.)

No. 31. A small kerosene lamp of such size and shape as to fit it for the use shown in Fig. 84. This figure shows the lower part of the chimney coated with asphaltum varnish, which is scraped off in one cross-shaped spot on a level with the bright part of the flame. A far better device is to surround the lower part of the chimney with a thin sheet of asbestos paper, having a hole 3 or 4 mm. in diameter at the height of the flame. The lamp must be watched to see that the flame does not grow too tall.

No. 32. A wire, of the right size to fit into the narrow slot of No. 20, bent at a right angle, one arm about 6 cm. long, the other about 4 cm. (Fig. 86.)

ARTICLES USED BY THE TEACHER, BUT NOT
TO BE FURNISHED TO PUPILS.

No. I.   A gauge for testing pressure at various points and
in various directions in a jar of water.   In Fig. 91, w is a
wooden column about 25 cm. tall and 1.5 cm. square; b is a
brass box, with brass bottom about 1.7 cm. wide and 1 cm.
deep; m is a thin rubber membrane fastened across the

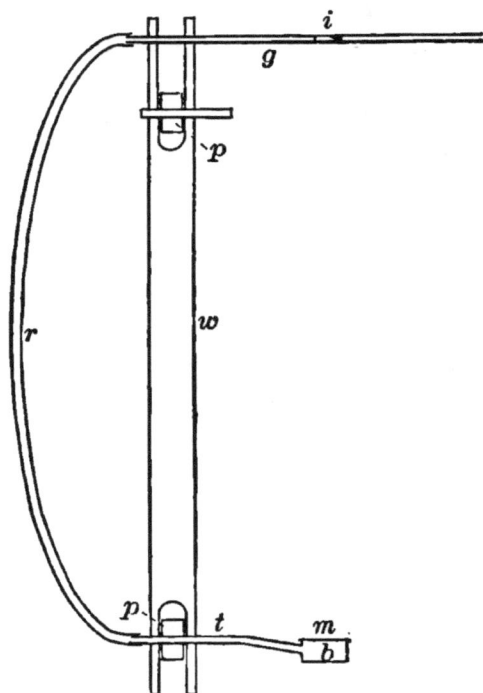

FIG. 91.

mouth of this box by means of sealing-wax—or, better, a
cement of pitch and gutta-percha; t is a brass tube about
8 cm. long and 0.5 cm. in outside diameter; p and p are
hard-rubber pulleys about 1.7 cm. in diameter fitting closely
on their axes; r is a small rubber tube; g is a glass tube

passing through w; i is a short column of water serving as an index. In use a band of strip-rubber, such as toy-stores supply, connects the two pulleys, p and p, so that by turning the axis of the upper pulley between the thumb and finger the gauge-face m may be turned upward, down-ward, or sidewise without changing level. A student-lamp chimney with stopper for one end accompanies this gauge.

No. II. Apparatus for bursting a bottle by an attempt to compress water within it.

No. III. Glass tube about 1 m. long, closed at one end, connected by a strong rubber tube 25 cm. long with an-other glass tube 20 cm. long. (See Fig. 10. In preparing this apparatus for use it is well to pour in mercury until the long glass tube *and half of the rubber tube* are filled.) A suitable support, with attached meter-rod, for holding this apparatus when used as a barometer, will be furnished with it.

No. IV. Strong *thistle-tube* (Fig. 11) about 2.5 cm. wide, covered at the mouth with strong sheet rubber and furnished with a thick-walled rubber tube about 20 cm. long.

No. V. Small air-pump suitable for both exhaustion and compression.

Only one experiment with the air-pump is described in this book, and for this experiment the pump alone, without base or plate for bell-jars, is sufficient. Manufacturers will supply a base and plate to be connected with the pump proper by means of a rubber tube, if it is ordered.

(With this air-pump and with the platform balance, No. XVII, and a large bottle, one may perform Exercise XI of Hall & Bergen's Physics, on the *Specific Gravity of Air*.)

No. VI. Bent glass tube for Boyle's Law, the whole tube about 1.5 m. long (Fig. 12).

No. VII. Common large rubber foot-ball, with a rubber tube about 30 cm. long attached to the key (Fig. 14).

No. VIII. Small bottle provided with rubber stopper fitted with two glass tubes as in Fig. 17.

No. IX. Glass model of lifting-pump (Fig. 20).

No. X. Glass model of force-pump (Fig. 21).

No. XI. Hydrometer for liquids less dense than water.

No. XII. Hydrometer for liquids more dense than water.

No. XIII. Glass U-tube (Fig. 22) about 60 cm. long before bending.

No. XIV. Lead Y-tube with attached rubber and glass tubes, and two small tumblers (Fig. 23).

No. XV. Eight-inch and four-inch wooden disks combined in one piece for use as a pulley. This piece is fitted with various pins (removable) for suspending weights. It is mounted much like the lever of No. 17. (See Figs. 26, 31, 32, 33, and 37.)

No. XVI. Centre-of-gravity board, with suspension and plummet (Fig. 27).

No. XVII. Platform balance weighing from 1 kgm. to 0.1 gm., provided with a set of brass weights.

No. XVIII. Well-made small brass pulley with a hook or loop (Fig. 39).

No. XIX. Well-made small double brass pulley with hook or loop (Fig. 40).

No. XX. An inclined plane * shown about one-fourth natural size in Fig. 92. The roller should be of brass, accurately turned. It weighs with its frame just 16 oz. The graduations of the scale may be in millimeters. The apparatus should be made with care.

No. XXI. Pendulum-support and pendulum-balls (Figs. 50 and 51).

No. XXII. Glass prism about 5 cm. long.

---

* A number of excellent features in this apparatus are due to Mr. Sweet of the Rindge Manual-Training School in Cambridge.

No. XXIII. Three small packages of dyestuffs soluble in water, various colors.

No. XXIV. Three glass plates, red, green, and blue, about 10 cm. square.

No. XXV. Camera obscura consisting of two paste-

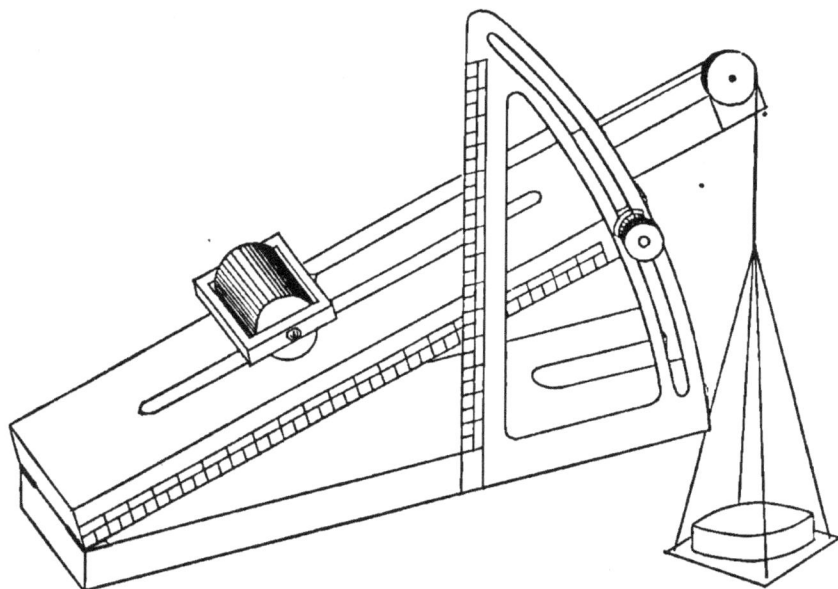

FIG. 92.

board tubes each about 25 cm. long. The larger, about 5 cm. in diameter, is closed at one end save at the centre, where there is a hole about 0.1 cm. in diameter in a thin partition. The smaller tube, about 4 cm. in diameter, is closed at one end by thin tracing-paper. (The observer pushes the smaller tube, closed end foremost, into the larger and then, pointing the apparatus toward a window, looks into the smaller tube and moves it back and forth in the other till the best image is obtained.)

No. XXVI. A concave spherical mirror 12 or 15 cm. in diameter.

MISCELLANEOUS ARTICLES.

Two pounds of clean mercury.

Two pounds of assorted soft glass tubing, from 2 mm. to 8 mm. inside diameter.

Six feet of rubber tubing, about 5 mm. inside, that will not collapse when connected with the air-pump.

Piece of thin sheet rubber about 6 inches square, for use with the gauge, No. I.

Set of cork-borers.

Three-cornered file for cutting glass tubing.

Screw-driver.

Pair of wire-cutting pliers.

One half pound of naked copper wire of the same size as article No. 32.

Small bottle containing a few ounces of mercury and an equal volume of chloroform, of water, and of kerosene (see p. 34).

COST OF APPARATUS.

Three firms of apparatus-makers, which in alphabetical order are

The Franklin Educational Company, Hamilton Place, Boston, Mass.,

E. S. Ritchie and Sons of Brookline, Mass.,

The Ziegler Electric Company (A. P. Gage & Sons), 141 Franklin St., Boston, Mass.,

have full information in regard to the apparatus mentioned in these lists, and will undertake to supply it.

The cost of a single set of the pupil's apparatus as described in the first of the lists, well made in every respect, will be not far from five dollars.

The cost of the apparatus in the second list and of the miscellaneous articles will probably be in the neighborhood of thirty dollars.

LABORATORY TABLES.

THE laboratory tables used in the Cambridge Grammar-schools are about 10 feet long, 4 feet wide, and 2 feet 10 inches tall. They have white-pine tops about $1\frac{1}{2}$ inches thick, and heavy white-wood legs. Extending from end to end over each table are two horizontal bars, about 2 inches by 3 inches, adjustable at various heights (which should range from $1\frac{1}{2}$ feet to $3\frac{1}{2}$ feet by 3-inch intervals) above the table-top, their ends, which are cut in tenons, sliding in grooves in the supporting posts. These posts are fastened to the frame of the table and rise through slots in the table-top, being *flush* with the ends of this top and about 10 inches distant from the sides. Pins of iron or wood placed in holes in these posts support the ends of the horizontal bars.

APPENDIX B.

By FREDERICK S. CUTTER, Master of the Peabody Grammar School
of Cambridge, Mass.

THE course of study for the grammar schools of Cambridge by a revision in 1892 was shortened and enriched. In shortening the course the work of six years was so arranged that pupils could complete it in five or in four years without omitting or repeating any part.

In the enrichment of the course one of the subjects added was elementary physics, by the laboratory method, and it was placed in the ninth or highest grade. The line of work to be pursued was laid out by Prof. Hall of Harvard University.

The time allotted to physics was one hour a week throughout the year, of which half an hour was for laboratory work and half an hour for recitation in the class-room. For the work in the laboratory the class was divided into divisions of 16 pupils or less. While one division was at work in the laboratory under an assistant teacher, the other pupils of the class were reciting or studying under the direction of the teacher of the class-room. Thus, for a class of 48, for a half-hour lesson in the laboratory, the assistant teacher used an hour and a half, while for a class of 56 or 60 two hours were required. In my own school, in which the class numbered 60 pupils, the following was the program:

116

Division.	2—2.30.	2.30—3.	3—3.30.	3.30—4.
I	Laboratory	Geometry	†Reading	Study
II	{ Geometry	*Laboratory	Study	{ Reading
III	{ Geometry	Study	Laboratory	{ Reading
IV	Study	*Geometry	†Reading	Laboratory

It will be seen that during each of the four periods two divisions together were taught by the teacher of the class-room, and one division was engaged in study. Thus the work in no way suffered from giving the laboratory instruction to small divisions. The half-hour for recitation was taken in a following session when all the divisions were taught together in the class-room by the teacher of physics.

With a class of 48 or less the following program could be used:

Division.	2—2.30.	2.30—3.	3—3.30.	3.30—4.
I	Laboratory	Study	Geometry	} Physics
II	Geometry	Laboratory	Study	recitation
III	Study	Geometry	Laboratory	}

The time for the teaching of physics—and also geometry, which has been introduced—was obtained in the revision of the program by completing the study of geography in the eighth grade, and by some modifications of the work in arithmetic. The one hour a week for physics was supplemented by making written accounts of the experiments a part of the language work. In the making of illustrations physics was further correlated with the work in drawing.

It was thought at first by some persons that a serious

---

* Recite together.   † Recite together.

objection to the introduction of laboratory physics into grammar schools, with their large classes, would be the amount of time and labor involved on the part of the teacher in preparation for, and in clearing up after, the laboratory lessons. But I have found that the teacher can be relieved of a large part of this labor by pupils selected from the class who will gladly serve as assistants. In the adjustment of the apparatus for the experiments nothing should be done for the pupils that they could properly be expected to do for themselves. But in taking from the cabinet, caring for, and putting away, the many articles used, the selected pupils can render valuable assistance. Thus, for example, one of my boys had charge of the 16 large glass jars,—the filling with water, the emptying, and the putting away in proper condition. Another pupil had the care of the 16 spring-balances; another, the overflow-cans; another, suitable strings and pins; etc. It was the duty of one pupil to see that everything needed for an experiment was finally in place, that there might be no needless loss of time on the assembling of a division. In the preparation for an experiment the names of the articles required were placed upon the blackboard, and the pupils having charge of these articles would see that they were rightly placed in the few moments before the opening of school, so that little or no time for this purpose would be required of the teacher. At the close of school the same pupils would see that everything was clean and dry, and put away in its proper place. The plan of giving to some pupils a share in the management of the work served to increase the general interest and to promote success.

INDEX.

Hall and Bergen's Text-book of Physics.

By EDWIN H. HALL, Assistant Professor of Physics in Harvard College, and JOSEPH Y. BERGEN, Jr., Junior Master in the English High School, Boston. xviii + 388 pp. 12mo.

This book contains the full text of the Harvard College pamphlet of experiments, interspersed with a considerable number of minor experiments and a large amount of discussion and problem work. The discussions have been carefully planned to enable the student to derive the full benefit of his experimental work and to guide him in his thinking; but not to relieve him of the necessity of thinking. Accordingly, wherever it has been found practicable, the conclusions to be drawn from an experiment have been withheld, and where they have to be made the basis of further work in the course, the statement of them has been deferred somewhat in order that the student may have opportunity to frame one for himself. The problems, which are very numerous, require the pupil to apply the knowledge he has gained from the experiments, and enable the teacher to ascertain how far the subject-matter has been mastered.

G. W. Krall, *Manual Training School, St. Louis:*—Is proving very satisfactory. It presents the true method of laboratory work in Physics. I have eighty students at work, and all are enjoying the experiments and taking far more than usual interest.

A. D. Gray, *William Penn Charter School, Philadelphia:*—My Physics class had already begun work when the book appeared, but the college preparatory division changed over to Hall & Bergen. I have been greatly pleased with the results, and only regret that I have not time to oversee the laboratory work of my entire class, done on the same plan.

Arthur O. Norton, *Illinois State Normal University:*—I find it more nearly adapted to our class work than any other work on the same plan that I have seen. The great trouble of works of this kind is that they tell the pupil too much. This objection is largely removed in Hall & Bergen's book.

J. H. Hutchinson, *Madison (Wis.) High School:*—Its introduction into our High School has been followed by excellent results thus far. I have taught Physics for some years, but never with so much satisfaction as during the present year. We now have about 100 using the book, and most of them are very much interested.

F. L. Sevenoak, *Stevens School, Hoboken, N. J. :*—We adopted some time ago as our method of teaching physics the plan upon which Hall & Bergen's book is based. No other plan gives such satisfactory results. I am glad to find a work so exactly suited to our needs.

Allen's Laboratory Exercises in Elementary Physics.

By CHARLES R. ALLEN, Instructor in the New Bedford, Mass., High School. *Pupils' Edition :* x + 209 pp._ 12mo.

Most of these experiments are quantitative. They are planned for young beginners, do not employ elaborate and expensive apparatus, require no more than forty-five minutes each in the laboratory, and are so framed that the pupil can prepare himself beforehand to make the most of his time there with the least help from his instructor. There are six exercises in Magnetism, eleven in Current Electricity, nine in Density and Specific Gravity, thirteen in Heat, twelve in Dynamics, four in Light, and two in Sound. Sufficient practice in methods of mensuration is also provided for. In the Teachers' Edition 67 pages are devoted to lists of apparatus with specifications for construction, suggestions as to substitutes, and itemized estimates of cost, together with detailed hints to teachers and references to standard text-books.

H. B. Davis, *Cushing Academy, Ashburnham, Mass.:*—We are using this year Allen's Laboratory Manual. I find it to be a book of the highest excellence. Especially is it noteworthy for clearness and conciseness, two of the most desirable qualities in a laboratory manual. With this book in his hand the pupil knows just what he is seeking after. He therefore knows just what to observe and what not to observe. It is a book without equal in my knowledge.

Edward Ellery, *Vermont Academy, Saxton's River:*—One of the most prominent features of the book, which recommended it to us particularly, is the plan of note-taking it suggests, a feature which I think is peculiar to this book. The experience of the term just closed has proved to our satisfaction that we were right in introducing it. We shall continue its use.

Arthur O. Norton, *Illinois State Normal University:*—It is unquestionably the best arrangement of quantitative work for beginners that I have yet seen. Am considering its adoption. [Adopted.]

J. C. Packard, *Brookline (Mass.) High School:*—At last we have a series of exercises that are not supplemented with conclusions and a list of questions that do not "contain their own answers." I shall introduce the book.

Geo. L. Chandler, *Newton (Mass.) High School:*—It seems to be an excellent book. I especially like the instructions for tabulating results. Whatever leads to systematic arrangement is a great help.

Chas. A. Mead, *Dearborn-Morgan School, Orange, N. J.:*—I am much pleased with it. The appendixes contain an amount of information that renders them invaluable to the teacher of physics.

Barker's Physics. ADVANCED COURSE.

By GEORGE F. BARKER, Professor in the University of Pennsylvania.
x + 902 pp. 8vo. (*American Science Series.*) *Revised.*

A comprehensive text-book, for higher college classics, rigorous in method, and thoroughgoing in its treatment of the subject as distinctively the science of energy. After a general introductory chapter, the subject is developed under three heads, *Mass-Physics, Molecular Physics*, and *Physics of the Ether.* Under mass-physics, energy is first treated of as a mass-condition, and work as being done whenever energy is transferred or transformed. The properties of matter are next considered, and then sound, regarded as a mass-vibration. Under molecular physics, heat alone is treated and as a manifestation of molecular kinetic energy. Under the head of æther-physics, which subject occupies three fifths of the volume, are grouped: (1) æther-vibration or radiation, considered broadly without special reference to light, (2) æther-stress or electrostatics, (3) æther-vortices or magnetism, and (4) ætherflow or electrokinetics.

"*The best truly modern manual of physics in our language,*" says the LONDON CHEMICAL NEWS.

Woodhull's First Course in Science.

By JOHN F. WOODHULL, Professor in the Teachers' College, New York City. In two companion-volumes.
I. Book of Experiments. xiv + 79 pp. 8vo.

II. Text-Book. xv + 133 pp. 12mo. Cloth.

One solution of the problem of very elementary science teaching—a solution that proceeds on the intensive rather than the extensive basis. These lessons make ample provision for a year's work, one period a week; they are entirely within the powers of average pupils ten or twelve years of age. All experiments can be performed on the pupil's own desk, without darkening the room. The necessary apparatus costs but $1.50 for each pupil, and most of it is in the nature of a permanent equipment.

Light has been chosen as the subject of these lessons because it exhibits a large number of phenomena at once capable of easy experimental development, and having numerous useful daily applications. The study is restricted to one branch of physics, partly to inculcate that thoroughness which is essential to true science, and partly for the sake of untrained teachers, who might be embarrassed by a wider range. Most of the work is quantitative in character, and in all of it each pupil must proceed independently and can be held strictly to account. Each experiment illustrates one truth, and is made the basis of exercises which correlate the teachings of the experiment with the pupil's daily experience and supply problems which connect his science most intimately and helpfully with his mathematics. The exercises are so arranged that the teacher can make the necessary adjustments to the varying degrees of aptitude in his pupils.

The material of each lesson is separated along its natural cleavage into that which gives directions and asks questions, and that which formulates results and contributes additional information when the pupil is in a position to desire and appreciate it. These divisions are put into the two mutually supplementary volumes. The Book of Experiments is also a note book.

Jackman's Nature Study for the Common Schools.

By WILBUR S. JACKMAN, Teacher of Natural Science, Cook County Normal School. Chicago, Ill. x + 448 pp. 12mo.

A practicable programme of lessons, comprehending the whole circle of natural objects and phenomena with which the child comes in contact. His equal and absorbing interest in every detail of his environment, in river, cloud, sunbeam, mountain, physical and chemical changes, and especially in living things as they live, is brought under control and directed to the systematic study both of the facts and their relations. It has been found that the power of continuous observation is thus cultivated and the habit of trying to account for things established. The pupil is directed through each topic by a series of questions and suggestions which he is to use in examining the things themselves, sometimes in their outdoor surroundings, sometimes by means of class-room experiments, and he is made regularly to record the results.

Evanston (Ill.) Course of Study:—Jackman's Nature Study is the best book published for the use of teachers. It is full of valuable suggestions for every month of the year in the several divisions of natural science, and each teacher may select therefrom such lessons as her judgment dictates.

C. Henry Kain, *Ass't Sup't Philadelphia (Pa.) Public Schools :*—I most enthusiastically endorse it. It is a handbook which no live teacher can afford to dispense with.

W. N. Hailmann, *Sup't of La Porte (Ind.) Schools :*—It is an excellent book compiled with much love for children and for the subject of instruction. The questions which are to lead to observation are wisely selected and clearly stated. The book offers the best solution I have seen of the problem of Nature Study in Elementary Schools.

New York School Journal:—Teachers will find by examining the book that it is possible to gain much of the science knowledge necessary by actual work; also that science teaching may be introduced successfully in any school and that materials in abundance may be found everywhere. The book is full of thought and helpful hints, and will give a great impetus to the study of science in the common schools.

George H. Martin, *Supervisor of the Boston, Mass., Public Schools, in* SCHOOL AND COLLEGE:—Each set of questions is preceded by a few practical suggestions, most of them eminently wise and helpful. It is in these questions that the teacher will find the book most useful. They teach him just what he needs to know,—what to look for in nature, how to read the "manuscript of God." What the teacher has found he can lead his pupils to find, and by reading the books to which the author refers, he can learn the scientific relations of his facts, and can answer the questions which he inspires.

Black and Carter's Natural History Lessons.

By GEORGE ASHTON BLACK, Ph.D.. and KATHLEEN CARTER. x + 98 pp. 12mo.

Part I. relates to the objects and operations that have engrossed man in his efforts to feed, clothe, and house himself. Part II. outlines a very elementary course in Botany and Zoology.

Bumpus's Laboratory Course in Invertebrate Zoology.

By HERMON C. BUMPUS, Professor in Brown University, Instructor at the Marine Biological Laboratory, Woods Holl, Mass. *Revised.* vi + 157 pp. 12mo.

The directions cover two representatives of the Protozoa, seven of the Cœlenterata, three of the Echinodermata, four of the Vermes, three of the Mollusca, seven of the Crustacea, two of the Limulus and Arachnoidea, three of the Atennata. An effort has been made to direct the work, without, at the same time, actually *telling* the student all that there is to be learned from the specimen. It is taken for granted that an instructor is present to assist when there is trouble, and to demonstrate many things that written descriptions might only render more confusing. In the Appendix a few words have been given regarding laboratory methods, etc.

M. B. Thomas, *Professor in Wabash College:*—I am very much pleased with it. It is clearly written and does not tell the student everything. So many manuals leave nothing for the student to find out for himself.

Albert A. Wright, *Professor in Oberlin College :*—It is admirable in its clearness of statement, admirable in what it omits. It brings into use a number of easily obtainable types which have not hitherto been treated of in laboratory guides.

Charles W. Dodge, *Professor in University of Rochester :* — One of the most valuable features of the book is the series of questions which the author judiciously uses along with statements which the student is expected to verify.

George Macloskie, *Professor in Princeton College :*—A book which I like very much and hope to adopt next September as text-book for a class in the subject which it covers.

Cairns's Manual of Quantitative Chemical Analysis.

By FREDERICK A. CAIRNS. *Revised and Edited by* E. WALLER, Instructor in Analytical Chemistry in School of Mines, Columbia College. viii + 279 pp. 8vo.

Hackel's The True Grasses.

Translated from " Die natürlichen Pflanzenfamilien " by F. LAMSON-SCRIBNER and EFFIE A. SOUTHWORTH. v + 228 pp. 8vo.

www.ingramcontent.com/pod-product-compliance
Lightning Source LLC
Chambersburg PA
CBHW021818190326
41518CB00007B/645